THAI FORESTRY

A CRITICAL HISTORY

THAI FORESTRY
A CRITICAL HISTORY

ANN DANAIYA USHER

SILKWORM BOOKS

ISBN: 978-974-9511-73-2

© 2009 Silkworm Books
All rights reserved

No part of this publication may be reproduced, stored in a retrieval system, or transmitted, in any form or by any means, electronic, mechanical, photocopying, recording or otherwise, without the prior permission in writing of the publisher.

Published in 2009 by
Silkworm Books
6 Sukkasem Road, T. Suthep
Chiang Mai 50200, Thailand
info@silkwormbooks.com
http://www.silkwormbooks.com

Back cover photo by Samphan Chayaraks Usher

Typeset in Garamond Premier Pro 11 pt. by Silk Type
Printed and bound in Thailand by O.S. Printing House, Bangkok

5 4 3 2 1

*For my parents, Samphan and Dan,
for Bjørn, Sasha, Luna, Sebastian, and Sindre,
and for Océane, Coco, Sabrina, and David*

A Non-Forester's Caveat

A number of wandering philosophers living near the Buddha had become noisy about their several views ... The All-Enlightened One ... listened and then gave them the parable. In former times a Raja sent for all the blind men in his capital and placed an elephant in their midst. One man felt the head of the elephant, another an ear, another a tusk, another the tuft of its tail. Asked to describe the elephant, one said that an elephant was a large pot, others that it was a winnowing fan, a ploughshare, or a besom. Thus each described the elephant as the part which he first touched, and the Raja was consumed with merriment. 'Thus,' said the Buddha, 'are those wanderers who, blind, unseeing, knowing not the truth, yet each maintain that it is thus and thus.'

Christmas Humphreys

ABOUT THIS BOOK

In the summer of 1987, I wrote a letter to the founding editor of *The Nation* newspaper in Bangkok, Suthichai Yoon, asking him for a job. At the time he was still working hands-on in the newsroom, and he called me in for an interview. I was without journalistic experience and my Thai was rusty. But what I lacked in qualifications I must have made up for in enthusiasm, and Suthichai agreed to put me on probation for three months. I stayed at *The Nation* for the next seven years. Articles written for the newspaper during that period are the starting point for this book.

This collection of articles and essays, so long in the works, is offered in gratitude to the countless people—villagers, activists, foresters, academics, and journalists—who taught, guided, inspired, and challenged me. I owe my first debt of thanks to Suthichai Yoon for giving me a chance, to Pana Janviroj, who encouraged me to join the paper in the first place, and to many other colleagues in the newspaper's editorial department. A grant from The Ford Foundation supported the research and writing. I have received additional small grants from the Margaret Lawrence Fund, the Grassroots Foundation, and Hatzfeldt Stiftung, and I spent two weeks at the Nordic Institute for Asian Studies in Copenhagen working on the manuscript.

I wish to express appreciation to the following people for different kinds of assistance over the years: Jannicke Bendixen, Raymond Bryant, Saneh Chamarik, James Charleton, Belinda Stewart Cox, Veerawat Dheeraprasart, Sanitsuda Ekachai, Charles Keyes, William Klausner, Pinkaew Laungaramsri, Siegfried Lewark, Karin Lindahl, Larry Lohmann, Robert Milne, Seub Nakhasathien, Kingkorn Narintarakul, Saskia Ozinga, Witoon Permpongsacharoen, Noel Rajesh, Y. S. Rao, Amnart and Napha Sanyong, Ildiko Schücking, Deirdre Shaw, Vandana Shiva, Chai and Pensri Siriwongwanich, Cecilie Steine, Nantiya Tangwisuttijit, and the staff at Silkworm Books.

I also thank my family, especially my parents, my husband, and our four children, for their support and patience.

A.D.U.

CONTENTS

Introduction 1

Part 1. Watersheds of Thai Forestry History 11
    Nam Chon 12
    The Logging Ban 17
    Huai Kaeo 22
    The Suan Kitti Scandal 25
    Khor Chor Kor 29
    The Death of an Honest Forester 32

Part 2. Scientific Forestry Enters Siam 37
    The Problem of Diversity: The German Forestry Model 39
    Herbert Slade's Legacy 51
    Colonial Strategies and Historical Resistance 66
    The Four Failures of Thai Forestry 73

Part 3. The Logical Conclusion: Factory Forests 89
    The Danish Factor 91
    Teak: Green Revolution Forestry 99
    Pine: A Window on Colonial Forestry 105
    Eucalyptus: Notorious *camaldulensis* 113
    A "New" Policy 123
    Industrial Strategies and Resistance 132
    The End of the Road 142

Part 4. The Making of Thai Wilderness 147
    Theft and Utopia: The American Model 148
    George Ruhle and the Story of Glacier 157
    Fathers of Thai Conservation 163
    Conservation Unlimited 173

Reinventing Thai Forestry 183

Notes 189
Bibliography 219
Index 231

INTRODUCTION

This book is about the thinking behind Thai state forestry. It is written by a non-forester—an outsider to the profession—and represents a challenge and plea to Thai foresters to look back on their own history with fresh eyes. It is not a book about Thai forests *per se* or the process of deforestation in Thailand *per se*, but rather an account of the historical ideas that influenced this process, and an examination of the intellectual roots of the style of forestry practiced in Thailand over the past century.

Thai forestry represents a strange mix of influences, land use models, and approaches to silviculture and conservation that were imported from Europe and North America during various historical periods. Each one carries with it the concerns, political baggage, and biases of its day. Forestry establishments in many tropical countries share a similar story, being based on theories of forest use invented in completely different social and biological environments and imposed without consideration for local conditions. Thai forestry professor Somsak Sukwong goes as far as to say that "there is no such thing as tropical forestry." By extension, there is no such thing as Thai forestry.[1]

Thailand's state forestry tradition does not reflect the country's astounding, multi-faceted biological and cultural diversity. Thai forestry has concentrated almost exclusively on a few tree species, focusing overwhelmingly on extracting timber and keeping local people away from the forest. Through much of its history, Thai foresters have ignored other types of flora and the vast body of human knowledge associated with their uses. The cultural diversity that embodies this ecological knowledge is reflected in the number of different languages in the country. Contrary to the popular perception of Thailand as a monolingual country, it has some eighty spoken languages.[2] Of a population of more than 60 million, only one in six people speaks "Standard Thai" as a first language.[3]

Forest knowledge is multi-dimensional. But medicinal plants that form the physical foundation of Thai traditional medicine have never been of interest to Thai forestry. Neither have mushrooms and other plants eaten fresh or cooked; nor have fruits, spices, and oils; nor has the proud tradition of cotton and silk weaving based on dyes made from roots, bark, plants, and fruits of the forest. Likewise, forest farming systems—swidden (slash-and-burn) agriculture in the north, forest fruit orchards in the south, techniques and rules of community forest use, and animist belief systems and rituals related to forest protection—are non-issues for Thai forestry. Fire, used by communities for millennia throughout Southeast Asia as part of forest management systems, is viewed solely as a threat to forests to be eradicated by foresters. Notwithstanding the impression of a monolithic "Thai Buddhism," there exist diverse ordination lineages of monks with their own distinctive interpretations of monastic rules, many of these embedded in forests. The unique phenomenon of forest conservation monks has never figured in Thai state forestry.[4]

In other words, state forestry practiced in Thailand builds neither on the rich and varied ecosystems that make up Thai forests nor on the vast reservoir of human knowledge possessed by forest-dependent cultures in the many ecosystems that make up the Thai forests. Instead, it has actively sought to remake and replace the forest with uniform ecosystems. Coming up against forest-dependent people, it has pursued policies that criminalize communities, leaving them with little option but to protest and resist.

These protests, initially sporadic, cautious, and unconnected, gained momentum in the 1980s and, over the course of the past two decades, have turned into a political force that fundamentally challenges the hegemony of the Thai forestry establishment. The most recent, concrete expression of the state being forced to respond to this public pressure is the passing of the Community Forest Bill, which gives legal recognition to the rights of communities to use and manage forest resources. The bill, though viewed as critically flawed by the popular movement that lobbied for its enactment, is nevertheless a fundamental challenge to both state bureaucratic power as well as the knowledge and beliefs that lie at the heart of Thai state forestry. The demand for community forests is a demand for land tenure in forest lands and, as such, directly contests the exclusivity of the forestry establishment's jurisdiction. Community forestry represents a different way of protecting forests—a democratized system of forest management—that undermines the state's model of conservation and foresters' accusations that poor farmers destroy forests. Moreover, to the extent that the community forestry movement has grown out of popular resistance to fast-growing tree plantations, the demand for community rights to manage forests takes on the ecological and social impacts of monocultures, debunking the official definition of "reforestation" and presenting in its place an alternative way of restoring forest ecosystems.

Community Forest Bill

Predictably, the Community Forest Bill that was enacted by the National Legislative Assembly on November 21, 2007, does not go as far as the popular movement that lobbied for its enactment wanted. Nevertheless, it represents a historic challenge to the forestry establishment, for the first time legalizing communities' rights to use and manage forests. The initial call by NGOs to grant community forest rights in 1989 marked the first rally in a political struggle that would continue for two decades. The struggle took the form of a debate over the provisions and content of a Community Forest Bill with versions of the bill proposed by the "people" countered by the Forestry Department's own versions.[5] While there was agreement in principle that communities should have some role in managing forests, as conservation began to dominate the state forestry agenda the Forestry Department became increasingly reluctant to relinquish control over newly gazetted lands, and disagreement concentrated on local people's access to protected areas.

After sustained farmers' protests in Bangkok—a full month in March 1996 and ninety-nine long days between January and April 1997—the government seemed close to conceding, but each time the demand for community access and control of forests inside the boundaries of protected areas lost out to arguments in favor of exclusive state jurisdiction in national parks and wildlife sanctuaries.

The community forest movement gained momentum after the passing of the 1997 Constitution. An exciting process of nationwide public hearings and broad public participation accompanied the drafting of the Constitution, which recognized more rights and freedoms than any other previous Constitution. It aimed at fundamental reform, including more transparency, accountability of elected politicians and public officials, and increased civil liberties. Direct citizen participation in the political process was facilitated through a provision whereby fifty thousand electors could submit a piece of legislation to Parliament. Secretary of the constitutional drafting committee Borwornsak Uwanno saw the new Constitution as offering, in particular, the possibility for NGOs and local interest groups to gain greater control over natural resources and the environment.[6] Clause 46 reads: "Individuals that together form an original (*dang doem*) local community have the right to preserve and revitalize their traditions, their local knowledge, their arts, and culture . . . and may participate in the management, recovery, and use of the environment and natural resources in a balanced and sustainable manner, in as far as the law permits." The community forest movement was the first within Thai society to put the new Constitution to the test and, in 2000, submitted a draft Community Forest Bill along with fifty thousand signatures. Though the bill again languished, the number of community forests in the national network continued to grow. By 2004, NGOs had documented over one thousand functioning community forests in the upper northern provinces alone.[7] Four years on, Somying Soontornwong at the Regional Community Forestry Training Center for Asia and the Pacific (RECOFTC) estimates conservatively that the real number exceeds fifteen thousand nationwide.[8]

The version of the bill that was finally passed in 2007 grants important concessions to communities, forcing the state to surrender control over forest lands in key areas.[9] Rights include access, use, and management of forests in national forest reserves, as well as inside protected areas, on the condition that communities can prove residence for at least ten years. Critics are not satisfied. They worry that in the absence of a registration system, verifying proof of settlement and actual management of community forests will be nearly impossible for ethnic minorities living in protected areas, some of whom lack even Thai citizenship. The bill forbids timber extraction from community forests, a provision designed to prevent a loophole for illegal logging, but one which critics say curtails sustainable forest use. The most serious problem with the bill, however, according to Chiang Mai-based activist Decho Chayathap of the Sustainable Development Foundation, is that in cases where a national park boundary separates a village from its community forest, people will not be granted community forest rights because this would entail "outsiders" entering the park. This means that a significant number of forest-dependent communities located outside of protected areas (likely some three to five thousand) face the dilemma of no longer being able to access resources on which their livelihoods have for many years depended. NGO workers like Decho argue that in its current form, the bill is so deficient that it breaches the 2007 Constitution's provision that recognizes community rights in natural resources management. NGOs have submitted a petition to the Constitutional Court. They are hoping the court will rule that the bill is unconstitutional and either remove the offending articles or throw it out completely. The court accepted the petition for review but, as of January 2009, no decision had been made as to the constitutionality of the bill.

The supremacy that the Royal Forestry Department (RFD) once enjoyed can hardly be overstated. Throughout the twentieth century, RFD was the single state agency responsible for protecting and managing Thailand's forest lands. From its establishment in 1896, it not only became the overseer of a key foreign exchange earner, teak, but was also an institution that facilitated the expansion and consolidation of the country's geographical borders through centralization of control over logging in the north. A governmental department with considerable status, RFD had by the 1980s a mandate to administer almost half the surface area of the country.

The demise of RFD has been a gradual process, fueled ultimately by the fact that the political power of the Thai forestry establishment was never matched by an ability to manage and protect the forests, even according to its own definitions and criteria. The proportion of the country under forest declined from approximately 75 percent in 1900 to 53 percent in 1960 to around 25 percent in 1998, according to official estimates.[10] During the years of most rapid deforestation following the Second World War, forest cover was disappearing at an annual rate of 4,500 sq km, among the highest rates in the tropical world. This coincided with the dawn of the development era and the Cold War, and resulted from a whole range of complex factors. Foresters were at best helpless bystanders and, at worst, corrupt colluders in this relentless process of denudation. Yet they did not question their own role in the tragic loss of one of the country's most valuable resources. As a profession, foresters systematically overlooked the manifold causes of deforestation and directed blame at one single group— rural people, especially ethnic hill people. By the latter years of the Forestry Department's century, with less than 20 percent (by unofficial estimates) of the country covered with healthy forest, criticism of forestry practices grew, and the forestry establishment began losing its political grip on forest lands. As forest cover continued to diminish, non-foresters in villages, universities, environmental organizations, and the media as well as among the general public came increasingly to question all aspects of the foresters' mandate: their monopoly over forest lands, their knowledge focused narrowly as it was on timber, their conspicuous loyalty to powerful private firms, and their consistent antagonism toward local people.

RFD's century ended officially on September 20, 2002, when the Thai Senate voted 127 to 23 in favor of separating activities related to commercial forestry and conservation. The result was that the Forestry Department was split into four separate entities. The most powerful of these was to be the National Parks, Wildlife and Plant Conservation Department under the newly formed Environment Ministry, which took with it much of RFD's staff and geographic territory. Two departments had responsibility for policy and coastal marine issues, respectively. Last and least was the Royal Forestry Department, which retained the old name, but with a staff and budget that was slashed by almost 90 percent. As part of the restructuring, the Forest Management Division, responsible for preparing timber-harvesting plans, was also dissolved entirely. The "new" RFD would have responsibility for the remaining forests outside the protected areas system—less than half the area it once commanded.[11] And as national parks and wildlife sanctuaries were designated in rapid succession, the National Parks Department would continue to expand its control, eating away at what was left of RFD's jurisdiction. The result was that the Forestry Department was reduced to a lackluster body lacking the mandate, prestige, and funding that it had enjoyed for the better part of a century.

This dramatic bureaucratic restructuring suggests a radical change in focus and direction, from exploitation to preservation. True, conservation is now the primary function of the Thai forestry establishment. But no new policy, law, or attitude accompanied the dissolution of the old RFD. Rather, the traditional dichotomy—timber production versus conservation—has remained. The contradictions between the two are as absolute as before, if not more so, and the social conflicts surrounding each are not resolvable within the current legal, political framework. The Community Forest Bill is, I would argue, in spite of its obvious shortcomings, the most innovative piece of Thai state forestry legislation to be passed in a century, since it resulted not from industry pressure or bureaucratic interests but from the demands of an unprecedented popular movement. It remains to be seen whether it will survive scrutiny by the Constitutional Court and, if it does, what its impact will be.

This book is organized into four main parts. It begins with a description of six "watersheds" of Thai forest politics. A watershed is a place in the landscape that indicates a border between two catchment areas—a point beyond which the landscape fundamentally changes shape. We can imagine a long climb uphill, finally reaching a place, a ridge, where the contours of the land begin to slope downwards again. The events described in this section marked historical moments—points along this ridge—that signaled a change in the political landscape, resulting from an emerging popular critique of state forestry management practices. For people who have followed Thai forest politics closely these past years, there won't be much new here, except perhaps the historical perspective; the sense that these events reflected a direct challenge to the imported silviculture and conservation ideas on which Thai state forestry has from its beginning been based. For those less familiar with the Thai forest story, the stories of these watersheds may provide some context. Each of the conflicts between the bureaucracy and the public contributed to the gradual erosion of the forestry establishment's legitimacy that continues to this day.

Part 2 of the book looks at three models of forest use that have most strongly influenced Thai state forestry policy, thinking, and practice. I begin chronologically with the scientific forestry of the colonial period, during which a British forester, Herbert Slade, introduced laws, a bureaucracy, and an education system that laid the foundations for the Thai forestry establishment. The type of forestry that he brought to Thailand was, as is the case in many other tropical countries, deeply rooted in the German forestry tradition, which sought to impose a strict grid of even-aged "normal forests" on wild, unruly nature. There was no room for people inside this vision, and that which lay outside the grid—whether other flora, animal life, or human communities—was quite explicitly viewed as waste. A crucial point here is that colonial foresters in Asia were well aware of the human presence in forest areas. This needs to be emphasized because popular wisdom tends simplistically to link deforestation and population pressure, explaining the antagonism between foresters and local people as a necessary response to growing numbers of poor. This is clearly not the case. In the early twentieth century, British foresters knew that "the many races inhabiting Siam [lived] in, on, and by the forests."[12] Claiming they had superior knowledge of forest management, early colonial foresters declared forests state property *because* of this human presence. Their purpose was to break the connection that human communities had with forests.

The promise of scientific forestry was that it would rationalize timber extraction in such a way that production could continue indefinitely.

This was never achieved in Thailand. More crucially, in Germany, where the science was invented, the long-term ecological impacts of intensive, single-dimensional forestry did not become apparent until the late twentieth century. Today in Germany and many other European countries, this approach is being abandoned in favor of a more holistic, ecological forestry.

The next part of the book views modern industrial plantation forestry as a logical conclusion of the narrow focus on timber species of colonial, scientific forestry. Thai forests have about three thousand tree species, but for the better part of a century Thai state forestry focused almost entirely on no more than three—teak, pine, and eucalyptus. In the key period after the Second World War when British and other European timber companies lost their logging concessions, Danish science and Danish aid money, oddly enough, played a key role in narrowing the focus of Thai forestry to these three species. Thailand's northern teak forests became something of a laboratory for applying to trees the green revolution theories that had made wheat and corn grow bigger and faster. Through this work, species of pine and eucalyptus came to be identified as having the highest potential for industrial plantations. Today, throughout Europe and in North America, the environmental consequences of pure factory forestry are increasingly questioned. "Are there really areas where we don't want to protect the groundwater and the flora and fauna?" asks a modern Danish forester.[13]

In Thailand, there is an added social dimension. Indeed, there are few areas of rural development that have generated as much controversy as eucalyptus. Communities' protests have centered on both the ecological impacts of this fast-growing exotic tree species and the land tenure and displacement issues that have often accompanied plantation projects. But Thai foresters have offered little in the way of guidelines or standards that would limit the negative impacts of plantations. They have more typically gone out of their way to defend industrial plantation forestry and to discredit critical questions and questioners. Eucalyptus continues to spread, particularly in eastern Thailand in the absence of solid data on the long-term impacts on soils and hydrological cycles of industrial tree farms. As such, it remains a risky experiment on a national scale.

If plantation forestry is a logical extension of colonial "sustained yield" logging, then conservation forestry is its mirror opposite. Part 4 looks at the introduction of the wilderness concept into Thai forestry discourse, which resulted from the unorthodox collaboration between a doctor, a dictator, and a well-intentioned nature guide from Montana. Using American national parks as prototypes, Thai forest conservation laws provided for the creation of large un-peopled nature reserves. In 1960 the wildcat notion of setting aside large areas of forests for nature conservation was viewed by Thai foresters as farcical. In the end though, with logging cancelled and plantations stalled, conservation achieved a dominant position in the Thai forestry establishment. But the expansion of national parks has been accompanied by relentless conflicts, since these areas are inhabited by millions of people, many from minority ethnic groups. It is a common, mistaken assumption that these conflicts are the result of a failure to apply properly the principles of conservation. In fact, the "prototypes" of conservation—Yosemite, Yellowstone, and Glacier—were not empty, but *emptied* landscapes. And many of the most spectacular American national parks, the so-called "crown jewels," are contested land up to today. So-called uninhabited wilderness, on which Thai conservation law is based, is nothing more or less than a myth.

The book ends with a question and a challenge: Western forestry, on which Thai state forestry is wholly based, has in recent decades begun the process of reinventing itself. In the Nordic countries, Germany, and North America, foresters are questioning their education and beliefs and rewriting the rule books on how forests should be managed. In the conservation world old ideas die hard, but also here the "Fort Knox" approach to biodiversity protection has given way to a more conciliatory approach to local people; realizing that accepting human presence in forest lands may not only be necessary from an ethical point of view, but also from an ecological one.[14] Where does this leave Thai state forestry? Criticism of Thai forestry has raged outside the profession, but few Thai foresters—and there are exceptions—have taken serious steps to reexamine the premises of their own profession. It is my hope that this book can stimulate a discussion among foresters about their own history and a possible future vision.

PART 1

WATERSHEDS OF THAI FORESTRY HISTORY

It is all about power.

Veerawat Dheeraprasart[1]

The history of Thai forest politics is marked by a series of challenges to the hegemony of the Thai forestry establishment. Each of the six "watersheds" of Thai forestry history presented here represented a crack in forestry's legal edifice—a fundamental questioning from outside the profession of the forester's worldview. These events, which I covered for *The Nation* during the late 1980s and early 1990s, serve not only as the starting point of this book but are the reason for the questions raised throughout these pages: Given that Thailand is a country whose forests are rich in both biological and cultural diversity, what can explain Thai foresters' single-minded obsession with a few species of timber and their seemingly desperate antagonism toward local people? What are the intellectual roots of Thai forestry? Why do Thai foresters think and act the way they do? Why did Thai state forestry fail?

The gigantic Nam Chon (Nam Choan) hydroelectric project, the first of the six "watersheds" discussed below, became the first major state project to be shelved due to public environmental concern. The area where the dam was to be built was subsequently listed as the country's first UNESCO World Heritage Site, and foresters who had argued publicly in favor of logging teak from the reservoir area were left standing like the naked emperor. The cancellation of logging concessions the following year stripped foresters of their original purpose, and dealt the Thai forestry administration a blow from which it has never really recovered. Soon after, the declaration of the country's first state-recognized "community forest" at Huai Kaeo (Huay Kaew) set a powerful precedent that directly challenged a century of centralized control over forest management. This launched a years-long debate over how to have these community rights recognized in law. The 1990 arrests of eucalyptus plantation workers for clearing away natural forest in order to make room for industrial tree farms—in the name of reforestation—exposed the fundamental ecological weakness in Thai reforestation policy. The resettlement scheme *Khor Chor Kor*— carried out by armed soldiers backed by forestry advisors—was the most extreme expression of misanthropic forestry. This plan to forcibly remove millions of landless people in order to free up space to plant trees was eventually revoked due to huge farmers' protests. Mass resettlement on this scale has not been tried since.

I have included as a sixth "watershed" of Thai forestry history an account of the struggles and eventual suicide of the Thai conservation forester Seub Nakhasathien. Seub's death pointed toward a new identity for Thai foresters: guardians of biological diversity. But it is surely at least as important that his life symbolized the frustrations and heroism of one person's fight against corruption in the forestry establishment.

Nam Chon

The Nam Chon dam was to be the largest hydroelectric dam in Thailand, 187 meters high with a generating capacity of 580 megawatts. It was to be built inside the Thung Yai Naresuan Wildlife Sanctuary, a 5,775 sq km area that forms the core of Southeast Asia's largest region of contiguous forest. Lying at the confluence of three major bio-geographic zones—the Sino-Himalayan, Sundaic, and Indo-Chinese—the area is rich in biological diversity. Studies showed that the vast range of faunal species in Thung Yai included 415 bird species, 82 mammals (such as elephant, tiger, gaur, wild buffalo, wild red cattle, leopard, deer, barking deer, Sumatran rhinoceros, and civit cat), 89 reptile species, and 52 fish species.[2] The forest had also been inhabited by Karen communities loyal to the Siamese kings for at least two centuries.

On March 18, 1988, plans for Nam Chon were shelved indefinitely by the cabinet in response to overwhelming public opposition. The title of *The Nation*'s page-one editorial that morning was "Stop the dam.

Listen to the people." The debate over the project had spanned two decades with concerns ranging from the political to the ecological. But one question had pivotal importance for the fate of the project: *Would the dam damage good forest*? Predictably, the Electricity Generating Authority of Thailand (EGAT), which was responsible for the project, went to some trouble to demonstrate that it would not. To prove this claim, EGAT enlisted the help of two leading Thai forestry professors who determined that there was nothing worth saving inside the flood zone. It became clear in the end that the project, in fact, threatened an area rich in biodiversity. And the support lent by foresters to EGAT's dubious claims seriously damaged public confidence in the credibility of the Thai forestry establishment.

The folly of Nam Chon was underscored when in 1991 forest surrounding the area that would have been flooded by the reservoir—the Thung Yai Naresuan and Huai Kha Khaeng Wildlife Sanctuaries—were added to the UNESCO list of World Heritage Sites because of its unrivalled biological diversity.

The idea of Nam Chon was first proposed in 1966. In the years that followed, EGAT tried again and again to obtain approval and the project was studied by several parliamentary committees and working groups. In the late 1970s, with the first two dams in the Klong river basin well underway, EGAT commissioned an environmental impact assessment of Nam Chon in accordance with National Environment Board requirements. The private Thai consultancy firm Team Consulting Co. Ltd. was hired, and Team, in turn, sub-contracted two Kasetsart University forestry professors, Choompol Ngamponsi and Kasem Chunkao, to design and carry out the forestry section of the survey.

The result of this effort was a report published in 1980, entitled *Environmental and Ecological Investigation*. It contained two main conclusions. First, the authors estimated that 10 percent of the 146 sq km of forest that would be flooded by the reservoir were already degraded due to farms cultivated by the Karen. Second, Choompol and Kasem compared the market value of clearing the entire flood zone with "sustained harvesting" of timber over the fifty-year project life, and determined that clear-cutting would be far more lucrative. They wrote:

"The total benefit from entirely clearing the Nam Chon reservoir area... is estimated at... 265.1 million baht *from teak*" (emphasis added).[3] The commercial value of all wood in the reservoir basin—calculated from the volume of wood and the market prices of the various types—was estimated at 639.7 million baht. In contrast, the net income of logging over fifty years would be only 159.7 million baht.

EGAT used these results to support its claim that Nam Chon would do little extra damage. Team's study was evidently good enough for the World Bank and the Japanese Overseas Economic Cooperation Fund, both of which indicated a willingness to provide financing should the project materialize. But though the project was put before the cabinet for approval in 1982, 1984, and 1986, it was turned down each time. It is likely that a series of earthquakes in April 1983 contributed to the delay. They were felt in Bangkok and their epicenters were reported to lie in the fault zone that runs under the reservoir of the Si Nakharin Dam, not far from the proposed site of Nam Chon.[4]

Moreover, at a time when the biological diversity of tropical forests was starting to become a favorite global *cause célèbre*, the exclusive focus of the 1980 survey on timber in one of Asia's most unique ecosystems caused a public outcry. A weakness of the study was, by the authors' own admission, that the survey work had mostly been conducted outside the actual flood zone. Communist insurgents controlled large parts of the western forests in 1980 and "more than 70 percent"[5] of the fieldwork was therefore carried out at sites presumed to have similar characteristics to those of the Upper Kwae Yai basin. Defending this choice of technique, coauthor Choompol said in an interview: "As long as you are examining the same type of forest it does not matter where you do it—a bamboo grove is the same everywhere."[6] But it did matter. The study was predicated on another more crucial falsehood: the foresters' claim that 265 million baht worth of teak could be harvested from the flood zone was absurd given that Thung Yai lies south of teak's natural range.

Even EGAT had to admit that the foresters' 1980 report had been "a little biased"[7] and the criticism was evidently serious enough to necessitate another review. The electricity agency hired the same two forestry professors to undertake a second study. But this time, it was not a full review of Nam Chon's potential environmental impacts. Rather, focusing on the key issue of public concern, they carried out a "forest inventory." Published in 1987, the new report concluded the following: The amount of forest degraded by local people in the reservoir flood area had increased since 1980 from 10 percent to almost 25 percent. The total amount of available timber had decreased slightly from 1.32 to 1.18 million cu m (cubic meters). But its total value had almost tripled during the seven years between the two studies to over 1.8 billion baht as a result of rising wood prices. Notably also, the 1980 reference to teak was absent in the new report.

EGAT now used this report as proof that Nam Chon promised to be environmentally benign. "Whether or not we build the dam, the forest is being destroyed by villagers bit by bit. We might as well build the dam, have the extra electricity, and protect what remains of Thung Yai that we love," said EGAT's public relations man.[8]

Surveys for the 1987 report were conducted inside the flood zone, which had by then become accessible. But it was otherwise an apology for the project, still focused on calculating the market value of clear-cutting timber and, apparently, on exaggerating the extent of forest degradation. Interviews with three Kasetsart University forestry students who worked on the survey team revealed how the 25 percent deforestation level was arrived at.[9] They claimed that the sample plots were not randomly selected but deliberately located in the vicinity of villages and in fallow fields. They also said that they intentionally avoided testing dense forest and steep areas, they worked for a total of only six days, and they tested only 227 plots (instead of the 400 indicated in the report). Moreover, if they were not able to identify tree species they were told to "guess" or give the Karen name.

The students also said that half the team was based in U-nite village in the north of the reservoir area, while the rest worked out of Mae Chan Ta, located near the center of the reservoir at the intersection of the Mae Chan and Mae Klong rivers. They slept in the villages, going out to survey during the day and returning before nightfall. This suggested that they were never more than a few hours walk from villages, and therefore more likely to be working in the vicinity of agricultural land and less likely to be testing relatively undisturbed forest.

Their Karen guides, selected by the Border Patrol Police from four affected villages—U-nite, Krueng Bo, Pai Nam, Mae Chan Ta—supported these claims. In interviews, they said they had offered to take the surveyors into "good forest," and were puzzled about why so much of the testing was carried out in old fields and areas with the smallest or fewest numbers of trees.[10] Both the students and the guides said no sample plots were located south of Mae Chan Ta village where there are no human settlements. If this is true, it indicates that the entire southern half of the reservoir area—where there was virtually no human activity—was not surveyed. It remains, however, impossible to know how representative the plots were as no maps were provided.

The forestry professors, Choompol Ngamponsi and Kasem Chunkao, told EGAT what it wanted to hear, and much of the forestry bureaucracy was in favor of the dam project. There were, however, foresters who raised critical questions. Sarayuth Boonyawechacheewin, for example, calculated the cost of replacing a deciduous dipterocarp forest of the type that Nam Chon would flood. He came up with the figure of 45 septillion baht—that is 45,000,000,000,000,000,000,000,000 baht, assuming, somewhat theoretically, that it would take about five hundred years to return such a forest to its original condition.[11]

On a somewhat more serious note, Phairoj Suwannakorn, who had been trained in national parks management in the United States, understood the value of Thung Yai beyond the market price of its timber. As director of the Wildlife Conservation Division at the time, Thung Yai came under his jurisdiction. When EGAT's workers began to cut a road to prepare for the dam, he ordered the regional forestry officers to arrest them. Though they were not yet inside Thung Yai Naresuan, they had already cut over 70 km, about half the total length of the road, 7 km of which were inside the adjacent Erawan National Park. "In fact, at the time, there was only an agreement in principle, approval had not been granted," he said in an interview. "If I hadn't stopped them, Nam Chon would be built by now."[12] There are many others who can rightfully make the same claim. But most of them are not foresters.

The Logging Ban

Not long after the Nam Chon dam project was shelved, public opposition to logging, which had been stirring for sometime, came to a head. In the weeks before the government's final announcement on logging, few people—certainly including those calling for the logging ban—believed that commercial timber concessions would ever be cancelled in Thailand. After almost a century, the Royal Forestry Department was practically synonymous with logging and the stakes, it seemed, were too high. There was no precedent in the region or, indeed, in the world, and powerful politicians were believed to have personal interests in the timber business. Most of the forestry bureaucracy was geared toward the management of timber extraction. Without logging, a whole branch of government might become redundant. Moreover, the Royal Forestry Department enjoyed a favored position in the state apparatus, whose very existence was inseparable from the history and geographical integrity of the modern Thai state.

Yet on January 18, 1989, the five-month-old government of Prime Minister Chatichai Choonhavan announced the end of Thailand's logging era. Reflecting on the decision years later, then Deputy Prime Minister Bhichai Rattakul said the decision was merely taken to increase the government's popularity.[13] A former timber merchant and a key figure behind the 1985 forestry policy, Bhichai said the ban had no scientific basis. It was a matter of "political survival," he said. Indeed, the country's first democratically elected government in twelve years cancelled logging concessions in response to overwhelming public pressure.

This pressure came from many different quarters. Opposition parties charged that as many as seven ministers had shares in timber companies,[14] students were threatening to sue the government for ecological damage caused by logging,[15] and even the influential elderly former prime minister M.R. Kukrit Pramoj urged that a logging ban would be "in accordance with the wishes of the public and might help the Government complete its four-year term."[16] Conservationists said the country's little remaining forest should be protected for values other than timber—for wildlife, for other biological diversity and genetic resources, for its role in maintaining climate and water supply, for its beauty and recreational values. While the official estimate of forest cover was 29 percent, it was widely believed that the real figure could be as little as half of this amount. If the principles of scientific forestry had functioned properly, and timber had been harvested "sustainably," it was argued, there should be more forest left to show for it.

Foresters defended scientific forestry, and instead placed the blame on the poor. Too many in number and lacking education, these villagers were the real culprits in Thailand's deforestation crisis, they said. This picture of reality had been the mantra of foresters from the time of the colonial forefathers—one that was also widely accepted by the urban public.

Nevertheless, a shift took place during the latter months of 1988. With Nam Chon fresh in public memory, and while the debate over logging raged in Bangkok, a different picture began to emerge from the countryside: one not of ignorant masses slashing and burning for subsistence, but rather of communities that had been struggling for years to defend patches of forest from state or private timber companies. With networking facilitated by NGOs, Project for Ecological Recovery in particular, protests became more frequent and better organized.[17] The demands of villagers in the north, east, and south to end logging in their local areas—demands that at any other historical moment might have been discounted by knowing experts and powerful bureaucrats—now added weight to prevailing sentiments.

By the 1980s, it had become apparent that little legally loggable forest remained in Thailand. Timber companies were facing increasing resistance from communities. They were also coming up against borders on all sides, borders with neighboring countries and borders with protected areas. An attempt by the Thai Plywood Company to regain rights to log inside the Huai Kha Khaeng Wildlife Sanctuary—adjacent to Thung Yai Naresuan, the site of Nam Chon—reflected the companies' desperation.

The state enterprise had been granted a 471-sq km timber concession in Uthai Thani Province in 1970. Ten years later the Huai Kha Khaeng forest, also in Uthai Thani, was given wildlife sanctuary status. And in 1986, two adjacent forest areas were annexed, almost doubling the total protected area to 2,575 sq km. The Wildlife Conservation Law forbids logging inside wildlife sanctuaries. The problem was that the two annexed areas overlapped part of Thai Plywood's concession.

Logging concessions were divided into thirty plots, one plot logged each year for a period of thirty years, theoretically allowing a concessionaire to return to the first plot in the thirty-first year to begin the process again. When Thai Plywood discovered that eighteen of its thirty plots lay inside Huai Kha Khaeng's boundaries, the state enterprise took the case to the Juridical Council.[18] The Council ruled that the 1961 wildlife law was not applicable in cases where concessions preceded the designation of sanctuaries.

The Thai conservation community was up in arms about the ruling. In October 1988, the provincial Uthai Thani Conservation Club collected ten thousand signatures during a local festival to protest the plan to log the sanctuary. In a very rare gesture of solidarity, some twenty Members of Parliament from both government and opposition parties also announced their opposition to the plan. With growing complaints from forest communities about the destructive actions of loggers, a meeting was organized by environmental organizations on November 26, 1988, at Kasetsart University to debate the future of Huai Kha Khaeng.

Meanwhile, far from the heat of Bangkok, a great natural disaster was unfolding in the southern provinces. Between November 20 and 23, more than a meter of rain fell on Surat Thani and Nakhon Si Thammarat. The rain quickly saturated the soil on the steep limestone mountains and, with few old forest roots to absorb it, triggered thousands of flash floods and mudslides.[19] Hundreds were killed, whole communities were literally washed away and billions of baht of property were damaged.

The rain made access difficult, and it took a few days before reports of the damage reached the national media. The glaring headline on the front page of the *Bangkok Post*, which appeared on the morning of the Kasetsart meeting, read: "Valley of Death."[20] Below these words was a photograph of a sea of uprooted trees, cut logs, and forest debris carpeting the valley floor. Widespread over-cutting, it appeared, had destroyed roots that might have held the soil to the steep slopes. Cut logs that had been carried down the mountainsides by torrential rivers of mud added to the damage below. By the end of the afternoon, the Kasetsart meeting's demand to protect Huai Kha Khaeng from logging had expanded to a general call for an end to timber concessions nationwide.

In the days that followed, public opinion crystallized. With local people pointing to the irresponsible practices of loggers, outrage over timber extraction inside a wildlife sanctuary, and horror at the effects of the floods in the south, the proponents of scientific forestry suddenly found themselves backed into a corner. A cartoon in *Siam Rath* newspaper ably described the atmosphere of moral and political persuasion that was brought to bear on a century-old institution. It pictured a writer holding a giant quill jammed under a log, trying to pry it loose with the force of his pen—the word of truth versus the weight of commercial interest. A drawing in the *Bangkok Post* depicted Prime Minister Chatichai marching forward, followed by a farmer, an academic, and a globe, chasing two loggers out of the forest. While canceling the concessions may have been a question of political survival, there was also a sense that it was in line with general public opinion, both in Thailand and internationally.

On December 8, 1988, forty villagers from eleven districts in six provinces—Chiang Mai, Phayao, Nan, Uthai Thani, Surat Thani, and Nakhon Si Thammarat—came to Bangkok. Along with members of the Student Federation of Thailand, they met with Prime Minister Chatichai at Government House to press for an end to commercial logging. Five days later, the cabinet banned logging in the twelve southern provinces. By January 17, 1989, the ban was extended to the rest of the country and on May 3, 1989, the ban took effect, officially ending Thailand's logging era.

Foresters had mixed reactions to the ban. On the one hand, there was resignation, and a sense that logging bans were the way of the future in Southeast Asia. Thiwa Saphakit, then director general of forestry, commented: "[The ban] is inevitable because there's so little natural forest left. By the year 2000, there will be logging bans in most of this region. We can't depend on a supply of timber from neighboring countries forever, we have to get serious about plantations."[21]

On the other hand, loyalty to the forestry discipline runs deep and foresters have shown little inclination to examine their own role in the tragedy of Thai forest loss. Ten years after the ban, the dean of forestry at Kasetsart University, Uthit Kud-In, still insisted that the government had made a mistake by canceling the logging concessions. "They treated [loggers] like children who have misbehaved at school, and forbade them to go to school ... Logging is not a problem in theory. Selective cutting when practiced sustainably gives you wildlife, biodiversity, beauty, and water. If you stop logging there are no pioneer species, only a climax forest. You might lose water because so many plants are demanding moisture. The public does not understand the principles of forestry. They just say it's impossible to control."[22]

The cancellation of timber concessions meant that the original purpose of the so-called National Forest Reserves—those areas that were "reserved" for logging under the exclusive jurisdiction of the Royal Forestry Department—suddenly fell away. By 1989 it was commonly estimated that some 10 million people (1.2 million families) were living illegally in these government lands. But the forest reserve category was not canceled along with the logging ban. Instead, these lands were targeted for commercial reforestation efforts to be undertaken by the private sector. The dilemma of what to do with these millions of "encroachers" became one of the most complex political legacies of the logging era.

If Bhichai Rattakul, former timber merchant and deputy prime minister at the time of the logging ban, could have had it his way, he would have turned a portion of these people into plantation workers and resettled the rest on 15-rai plots of tenured land in areas unfit for commercial tree farms.[23] But resettlement projects have proved problematic at best, and there has been widespread local resistance against plantations. Rural communities living without title in and around forests, especially in the north, northeast, and east, have fought against threats of eviction by offering to take care of "their" forests better than the plantation companies.

Huai Kaeo

Huai Kaeo was one of those many communities that tried to defend their forest against state-sponsored projects. Unlike every other community living illegally in government forest land however, the people of Huai Kaeo were granted provisional rights to manage and use the forest. The creation of Huai Kaeo Community Forest was a wholly unprecedented ruling whose legal and political implications were potentially enormous.

In 1986 Chiang Mai MP for the Chat Thai Party, Suraphan Shinawatra (an uncle of former Prime Minister Thaksin Shinawatra), became deputy communications minister. While holding this post, he lobbied for a number of projects in his province, one of which was a concession for 235 rai (about 40 hectares) of "degraded forest" in the Mae On Forest Reserve for the purpose of planting mango and other fruit trees.

The concession was issued in the name of Suraphan's wife, Prameun. It was granted in the spirit of the 1985 forestry policy, which encouraged the private sector to invest in tree-planting schemes to help the underfunded Forestry Department reforest the country. In February 1989 the cabinet approved the orchard project and almost immediately a controversy flared up. It centered on two questions: Do mango trees constitute reforestation? And was the forest area degraded in the first place? Foresters said "yes" to both. Huai Kaeo villagers said "no."

Though the Shinawatras vigorously denied they had damaged any trees or committed any other illegal acts, evidence inside and outside the concession area suggested otherwise. Local people claimed that tractors and heavy earth-moving equipment were sent in to prepare the ground for mango saplings. Their fears were compounded by unsubstantiated rumors that the orchard was simply a front for a tourist resort development project that would be constructed once the couple had consolidated control over the area. The villagers said workmen felled large old trees, and caused massive soil erosion by cutting an access road into the steep hillside. Evidence of both could be observed more than two years later.[24] A *mueang fai* traditional irrigation canal that was once two meters deep was now filled with earth. Fields had been clogged with sand. Logs buried by workmen underneath the soil had become visible, uncovered by a trickle of water where a stream had once flowed.

Led at first by a local schoolteacher Nit Chaivanna, Huai Kaeo villagers called for the expulsion of the developers and for "community rights" over the 1,600 rai of forest on which they depended. Today, with thousands of functioning self-declared community forests around the country, these demands do not seem extraordinary. But in 1989 the demand for community rights was groundbreaking. Kasetsart University forestry students surveyed the area and concluded that not only was the 235-rai forest not degraded, but also the concessionaire had encroached upon forest outside the permissible area. (The Shinawatras had originally requested the use of some 600 rai and were only granted 235 because the surrounding area was deemed to be good forest.) Two months later, the students' findings were confirmed by the National Counter Corruption Commission. It ruled that forestry officials had been "negligent" in allowing good forest to be rented out.

The Huai Kaeo controversy intensified as more Chiang Mai University students and academics joined the cause. Tension peaked on December 14, when Nit was arrested along with a young villager leader Thaweesin Srisaengnam and student activist Vichien Anprasert. Prameun Shinawatra had accused them of stealing her mango saplings, damaging her property, and encroaching on her concession land. Other villagers rallied around the jail, daring police to arrest them as well. With each development being reported in detail in the national press, the three were immediately released. But on the following day, the teacher's mysterious murder heightened the tension in the air among the villagers.

On December 18, 1989, Phairoj Suwannakorn, who had by then been appointed director general of forestry, made an emergency visit to Huai Kaeo. Under the scrutinizing eyes of villagers, students, academics, environmental activists, and the national media, Phairoj bowed to public pressure. He agreed to give Huai Kaeo villagers control over the 1,600-rai forest, overruling the legal claims of the influential Shinawatra family. The forest would be under the responsibility and control of local people for local, non-commercial uses, and a committee comprised of villagers, civil servants, non-governmental organizations, and academics would oversee the management. Though covering only a small area, the ruling marked a vital turning point in Thai forest politics. State recognition of a community forest in Huai Kaeo was, literally, a revolutionary departure from the centralized approach to forest management that had dominated the thinking until then. With no law to underwrite his ruling, the director general had to declare the community forest to be an "experimental project," whose legal status was to be determined.

Still, the precedent set by the creation of the Huai Kaeo Community Forest exposed weaknesses in existing policies that were encouraging destruction of intact forest in the name of reforestation. Moreover, Phairoj's action forced the government to grant official recognition to people living "illegally" inside forest reserves. Most significantly, local people had now been given virtually exclusive control over the resource. Never in the ninety-three-year history of the Forestry Department had a decision been taken that had the potential to weaken the agency's hold over the forests to such an extent.

The idea of legalizing community management of forests through a community forest bill was first proposed by NGOs in 1989 as a logical next step following the logging ban, and it gained momentum after the recognition of Huai Kaeo Community Forest. In subsequent years, the community forestry movement became national in scope, notably through association with the national network Assembly of the Poor, which came into being in 1995. Drawing inspiration from the 1997 Constitution—which recognized the rights of communities to manage local natural resources according to their cultural beliefs, and allowed for fifty thousand citizens to submit legislation to Parliament—a draft bill was proposed in 2000. Objecting primarily to allowing local people access to forests inside protected areas, the Forestry Department countered the proposal with its own version of the bill. This tit-for-tat dragged on for years with apparent victories and disappointing setbacks for proponents of community rights.

Finally, however, in November 2007, the military-appointed National Legislative Assembly passed a version of the Community Forest Bill that, while blocking villagers living outside national parks and wildlife sanctuaries from using forest in protected areas, grants a key concession. It recognizes the right of communities established inside parks and sanctuaries to manage their community forests if they can prove they have been doing so for the past ten years. It should be noted that the bill could still be revoked. The Northern Farmers' Network is challenging the constitutionality of the bill with the Constitutional Court. The NGO argues that denying non-residents of protected areas community forestry rights contradicts the spirit of the new 2007 Constitution, which, in spite of being written under a military government, is in its language even more progressive on the question of community natural resource management than its predecessor of a decade earlier. The NGO network also objects to the restriction of villagers' use to non-timber forest products, arguing that they should be permitted to cut timber as well, within agreed guidelines. Should the court rule in their favor, the entire bill could theoretically still be thrown out.

The Suan Kitti Scandal

Not a month after the Shinawatra's mango project was rejected another forestry scandal caught the attention of the Thai press. It was sparked by the sudden arrests in January 1990 of 156 workers of the Suan Kitti Reforestation Company in its eucalyptus plantations in the eastern province of Chachoengsao. The immediate charge was illegal clearing of natural forest to make way for fast-growing eucalyptus trees.

Like the Huai Kaeo controversy, the Suan Kitti scandal focused public attention on the destruction of natural forests caused by commercial reforestation. But while the former was a clear case of disregard for existing regulations, the latter illustrated that compliance with regulations could also lead to forest degradation. The political crisis surrounding the scandal intensified critical discussion of forestry policy, and eventually caused the cabinet in May 1990 to suspend the practice of renting out large areas of state land for tree plantations.

Kitti Damnerncharnvanich, the owner of Suan Kitti Reforestation Company, was a senator, prime ministerial advisor, patron of the Democrat Party, and possibly other political parties, and one of the most influential agro-business tycoons in the country. At different moments in his illustrious career, Kitti had been the world's largest exporter of rice, and then of tapioca. In the early 1980s, a trip to Brazil to visit the part Norwegian-owned pulp mill and plantation complex Aracruz Cellulose inspired Kitti to try his hand at eucalyptus.[25]

The Aracruz mill was supplied with wood chips from more than 100,000 hectares (625,000 rai) of eucalyptus planted and owned by the company. Kitti must have realized that since Thai law did not permit private ownership of forest land, it would be difficult to recreate exactly the conditions of Aracruz. Still, he set about to gain control of as much land as possible. He had three options: legal purchase of land from titled farmers, rental of public land from the Royal Forestry Department, and purchase of influence through paying untitled farmers to leave the land.

Ministry of Agriculture regulations placed a 2,000-rai (320-hectare) ceiling on degraded land rented by private companies, ostensibly to limit the size of holdings. Anything larger would require the cabinet's approval.[26] To get around this rule, Kitti issued dozens of requests to the Ministry of Agriculture for plantation concessions, each one just under 2,000 rai. He also stationed a company employee full time in the ministry to facilitate the movement of red tape. Obtaining permission from the minister required months of waiting—for forestry officials to survey an area to determine whether it was indeed "degraded," and if not, to adjust the proportions of the concession accordingly, and for appropriate documents to be signed and counter-signed. In addition to this official process, it was necessary to compensate small farmers living illegally in those areas to avoid protests or possible retaliation in the form of incendiarism or demonstrations.

Once permission was obtained, the land could then be prepared for planting. That involved removing all plant material from the area to create optimum conditions for eucalyptus. The larger trees were burned, buried, or removed, and the fields were plowed over three or four times using tractor and blade to loosen the soil and remove roots. Fertilizer was then applied before the eucalyptus saplings were finally planted.[27] But the key was not to begin this process before final permission was obtained from the minister.

When the arrests were made, the Suan Kitti workers were accused of operating on land where the company had not yet received official permission to plant. The public was up in arms about the ecological implications of a forestry policy that encouraged, even necessitated, the dismantling of the forest structure to make way for tree farms. In fact, strictly speaking, Suan Kitti's crime was not ecological but administrative. If the company had waited a few weeks longer before preparing the fields for planting, the operations would have been in perfect compliance with government reforestation policy.

Following the flurry of publicity, a group of reporters from *The Nation* were invited by the Damnerncharnvanich family to visit the controversial area in eastern Thailand. Kitti and his son Yothin acted as our guides. The tour was organized as a concession to the newspaper for allowing Kitti his full say in a page-one interview with Peter Mytri Ungphakorn, at a time when other Thai dailies were accusing the former senator of all manner of heinous crimes.

Lining both sides of the road leading to the Suan Kitti complex were vast stands of eucalyptus trees standing in orderly columns as far as the eye could see. Kitti told us that the last great challenge of his life was to establish a 400,000-rai (640-sq km) eucalyptus plantation to supply a 1,000-tons-per-day pulp mill and a self-sufficient Thai paper industry. He said he had already obtained about 25 percent of this goal with 100,000 rai of eucalypts planted in one hundred separate plots. He would secure the rest by a combination of straight purchase of land title and rent of Royal Forestry Department land. "If politics doesn't finish me off, I will succeed in this project. I've come too far to back down now," he told us.[28]

Inside the 1,000-rai complex, Kitti showed us through a long row of cylindrical greenhouses. In each one, tiny purplish eucalyptus seedlings were being cultivated, eventually to be added to the adjacent plantations. But it was in the "clone bank," as he called it, where selected trees were being tended individually, that Kitti's hopes evidently rested. Here, the fastest-growing specimens chosen from thousands of rai of plantations were being replanted, and their seeds and tissue studied in an on-site laboratory. Walking up to one particularly robust tree, Kitti wrapped his fingers around the trunk to show its impressive diameter. "This one's only three years old and already big enough to be chipped. If all the trees in my plantations could grow like this, we would be producing faster than the Brazilians!"

Kitti's success, like that of similar projects throughout the tropics, depended on his ability to breed an individual with optimum genetic characteristics that would correspond to the demands of the pulp industry. The ideal would be a straight, fast-growing tree with narrow canopy except for widely spreading branches close to the ground to stunt the growth of grasses that compete for limited water and soil nutrients. This high-yielding individual would then be mass produced and planted throughout the Suan Kitti plantations of eastern Thailand. Aracruz promotional material at Kitti's Chachoengsao office explained how a maximum consistent yield of high-quality pulp was to be ensured. "Multiplication is accomplished through rooted cuttings: cloning. The trees grown from each clone are genetically identical," the pamphlet stated, and a single word printed in large, bold letters summarized the spirit of forestry practiced in Asia since colonial times: "HOMOGENEITY."

Following the scandal, the government changed the regulations governing the rental of state forest land, decreasing the maximum leasable area from 2,000 rai to just 50 rai. Public concern about the ecological implications of eucalyptus tree farms intensified during this period, and Kitti Damnerncharnvanich faced a serious setback. He was, however, down but not out, as we shall see later.

Khor Chor Kor

Suan Kitti's plantations were not the only ones delayed, and government projections about how much land could be covered with eucalyptus trees proved overly optimistic. By 1987 the Ministry of Agriculture had identified 28,800 sq km of "degraded" land where eucalyptus could in theory be planted.[29] But only 2 percent—about 640 sq km[30]—were planted with the fast-growing trees. Four years later, the area covered with eucalyptus had increased to about 1,600 sq km,[31] but this was still far short of the ministry's ambitious target. Landless people living in state forests were seen as the main obstacle to the successful implementation of reforestation as defined by Thai forestry policy.

But where foresters had failed to keep people out of state forest lands, the army now stepped in. With the commencement of the Land Resettlement Program for the Poor Living in Degraded Forest Reserves—whose Thai acronym was *Khor Chor Kor*—state tree-planting efforts were stepped up and repression against landless people became more organized and systematic than ever before. Heavy-handed tactics traditionally used by foresters against local people assumed nightmarish proportions under this military resettlement scheme. It is barely conceivable in retrospect that state repression on such a scale could be planned and even partly implemented in Thailand at a time when other democratic freedoms were practically taken for granted. The most remarkable thing about *Khor Chor Kor*, however, is that it was eventually revoked as a direct result of political pressure from landless villagers.[32]

Designed with the assistance of the Forestry Department and approved during the Chatichai Choonhavan administration, widespread implementation of the scheme began after the military coup d'etat of February 1991. *Khor Chor Kor*'s stated purpose was to resolve the land issue in order to reach the national goal of 40 percent forest cover. The plan was to "resettle" 970,000 families (some 6 million people) living illegally inside National Forest Reserve land in orderly villages under military control. Villagers who were estimated to be using some 22,900 sq km of agricultural land would be squeezed onto 7,680 sq km, making available about 15,000 sq km for reforestation.[33] Though it was to be carried out eventually in all four regions, *Khor Chor Kor* began by targeting 250,000 families in the northeast for resettlement.

Each relocation site was supposed to accommodate one hundred families, and was to be arranged in one of three patterns—a ring shape, a square block, or a straight line. No apparent allowance was made for cultural and linguistic difference, nor attention paid to the appropriateness of the soil for cultivation. The result, in the words of activist Sisuwan Kuankachorn, was a "total mess."[34] In some cases, soldiers cleared away forest to make way for the newcomers. Often, the "new" land was already inhabited, causing impossible conflicts between the original owners and the newcomers. In still others, villagers, especially those who refused to cooperate, were given no land at all.

Communities that had formed over decades or centuries along kinship, ethnic, and linguistic lines, settling according to geographic particularities of each area, were literally "torn from their land."[35] Soldiers moved into villages, giving ultimata for inhabitants to pack up their belongings. In some instances, if people failed to cooperate, armed men took down houses and schools, destroyed temples, and even plowed up fields of crops to force people to leave. I visited Nong Yai village in Soeng Sang District, Nakhon Ratchasima Province in September 1991, the day after soldiers had laid waste to several hundred rai of tapioca fields. As we arrived, farmers were hurriedly salvaging the roots, as rifle-bearing soldiers looked on.[36]

The inhabitants of Nong Yai were among those whose months-long resistance played a crucial role in the eventual demise of *Khor Chor Kor*. They had been forced from their homes by soldiers a month before. The eviction received special attention at the time because of the involvement of Phra Prachak, a forest monk and founder of the controversial Wat Hua Nam Phut forest temple. He had been working with villagers in the area, patrolling forest for illegal loggers, and preaching about the Dhamma of nature protection. His violent arrest by police and soldiers earlier that month pushed Buddhist tolerance beyond its normal limits. A young woman, Nittaya Pamala, was traumatized by what she said was her first experience of this kind:

"I wish you could have seen it. Soldiers kicking and hitting people. The ones who tried to protect Phra Prachak were dragged away, beaten up, and thrown into the military trucks as if they were pigs or dogs—not human beings who live and breathe and eat rice like you and me!"[37]

Upon arrival at the relocation site, some discovered that they had no new land, while others found their allocated plot was already occupied. For months, more than a hundred Nong Yai families remained in limbo—landless, homeless, and uncertain about the future. Until June 1992 they clung together in tents, suffering from hunger, illness, and despair. This was the plight of thousands of other northeasterners who had experienced similar treatment. The people from Nong Yai were surrounded by soldiers and were too afraid to move back to their village. In any case, the village had been razed to the ground, and they had no obvious place to go. A cloth sign, hung outside the temporary shelter that served as home for nine months, read: "Thai people's refugee center."

Finally, fed up with waiting, about three hundred people returned to the original site of Nong Yai in mid-June, in open defiance of state authorities. One villager described her excitement: "Today, for better or for worse, we are going home. We are afraid. But if we don't move now, we won't have anything left. The military destroyed our homes and took our farmland away. For nine months, we have been waiting. But they could not fulfill their promises. It's a dead end. We have to take this risk, for we have nothing more to lose."[38]

The Nong Yai protest was one of many sweeping Isan. A week later, thousands of landless farmers from eleven northeastern provinces walked to Pak Chong in Nakhon Ratchasima Province, threatening to march on Bangkok if *Khor Chor Kor* was not revoked. Revulsion of the Thai military's bloody crackdown of peaceful pro-democracy demonstrators in Bangkok just a month prior boosted public sympathy for the villagers' demands. On July 3, 1992, the newly appointed Prime Minister Anand Panyarachun suspended the resettlement plan. *Bangkok Post* journalist Sanitsuda Ekachai commented afterwards that "the . . . defiance of the Nong Yai villagers against this *Khor Chor Kor* scheme has brought many hidden issues out in the open. It has shown that the bureaucracy is still the main alliance of the military to suppress democracy. It has also highlighted the danger of environmental destruction under military dictatorship."[39]

Among these "hidden issues" was the under belly of scientific forestry. *Khor Chor Kor* was a raw show of military force exercised in the name of protecting forests *for* the state *from* the people. While foresters had neither the capacity nor the mandate to use the tactics of soldiers to remove villagers forcibly from state-owned forest land, Forestry Department officials participated in the scheme, assisting with the planning, and offering advice about which areas should be cleared to make way for reforestation.[40]

Like the Huai Kaeo Community Forest, whose implementation has been marred by endless bureaucratic and political obstacles, *Khor Chor Kor*'s cancellation offered no definitive solution to Thailand's land and forest tenure conflicts. Its significance lies elsewhere. Firstly, the capacity of landless farmers to influence decision makers at such a high level set a powerful precedent in itself. But *Khor Chor Kor* also provided a kind of yardstick of the level of force that is politically tolerable. If Huai Kaeo, where farmers manage local forests, is at one end of the scale, the *Khor Chor Kor* military resettlement scheme, where soldiers threw people off their land at gunpoint to free up areas for tree farms, is at the other. To this day, Thai foresters maneuver between these two extremes: determined to steer away from the former, yet fully aware that the latter goes beyond what Thai society is willing to accept.

The Death of an Honest Forester

On Saturday, September 1, 1990, just before daybreak, the Chief of Huai Kha Khaeng Wildlife Sanctuary Seub Nakhasathien took his own life, using the hand pistol he carried with him for protection. Seub's death sent shock waves through the Thai conservation community, and literally moved the nation.[41]

The suicide differs from the other events described in this chapter. But it, nonetheless, is a "watershed" of Thai forestry history. The public's reaction to this forester's death reflected the loss of national patience with corruption and collusion within the forestry bureaucracy, and it heralded the coming shift of focus in Thailand from production to conservation forestry.

Seub was a totally different kind of forester. Accustomed to the stereotypic image of the corrupt timber monger, Seub had won people's hearts through his struggles to make publicly known the dire circumstances of Thailand's wildlife. They saw him rescuing animals left stranded by construction of the Khao Laem hydro dam, exposing illegal loggers at great personal risk, speaking out against the Nam Chon dam project, risking his life to protect helpless wildlife—in short, taking a stand where many of his contemporaries either actively colluded with illegal traders or chose to remain silent.

"I am speaking today on behalf of the animals in the forest because they cannot speak for themselves," he told a gathering of one thousand people at the Environment '90 Conference held in January 1990. Showing slides of caged baby primates and severed tapir hooves, he spoke directly from the heart. He was the only speaker at an otherwise ordinary conference for whom the audience stood up and cheered.

Seub's death nine months later provoked an exceptional outpouring of grief and affection from people in all walks of life. Newspapers were inundated with letters and poems. One of Thailand's most popular singers wrote a song in his honor: "Seub Nakhasathien was a lesson for the Forestry Department . . . a lesson for the Thai government."[42] A television fund-raising program collected over 10 million baht in one evening, which was used to set up a foundation in his name. The statue of Seub now stands at the Huai Kha Khaeng Sanctuary headquarters. In a letter written five days after his death, His Majesty the King made it known that he would sponsor the final two days of the seven-day funeral rites, sending his privy councilor and former prime minister Prem Tinsulanonda as his representative to the ceremony. In his "Soi Suan Phlu" column in *Siam Rath* newspaper, M.R. Kukrit Pramoj compared Seub to a soldier who died for his country in the line of duty. As soldiers are recognized for their deeds, he wrote, so Seub should be remembered as a national hero.

The analogy of war was not to be taken lightly. Seub often said his struggle to protect the integrity of Thailand's western forests was like a military conflict, except that it was being waged with none of the support systems necessary for those risking their lives on the front lines. It was not a fair fight. While the "enemy" was fully equipped with M-16s and other military weaponry, and benefited from political connections and external financial backing from Bangkok and beyond, the forest rangers of Huai Kha Khaeng were virtually on their own. Overseeing 2,575 sq km of forest were 12 officers, including Seub, 30 rangers, and 120 guards, some of whom were paid no more than 1,500 baht (less than US$50) per month with no benefits or insurance for their families in the event of death. On top of endless paper work and uninterested responses from officialdom to their pleas for assistance, these men had also to contend with malfunctioning radios, vehicles badly in need of repair, and a woefully inadequate fuel budget. It is in the hands of this sorry crew that the government placed full responsibility for one of Thailand's most precious remaining natural treasures.

It was largely as a result of the efforts of Seub and the British zoologist Belinda Stewart Cox that Huai Kha Khaeng and the adjacent Thung Yai Naresuan were recognized by UNESCO as a Natural World Heritage Site. Seub's concern for the survival of wildlife inside the sanctuary stemmed from a solid conviction that animals have rights to life as human beings do. He asserted that the world was not made by us or for us, nor does it belong to us. He believed that hunting wildlife for sale was morally equivalent to murder.

"People who buy baby animals [like gibbons and slow lorises] to raise as pets at home must take the responsibility all the way, not give them back to the Forestry Department when they become adults and start to bite, and are no longer cute. It's like having a daughter or son and giving it to the Social Welfare Department to raise," he said.

So many times Seub raged at how absurdly out of joint the government's priorities were. "When protected forest is damaged we get a budget of 1 baht per rai to restore it. But if it's forest reserve land they allocate 1,000 baht for planting eucalyptus—and then they call that forest," he said.

During the last months of his life, Seub became caught in a bureaucratic web. The more he struggled, the more he felt trapped; he was losing control of the things that meant the most to him. He found himself fighting against poachers in the villages surrounding the boundaries of the sanctuary, some of whom were family members of his own workers. But Seub was no misanthropic conservationist. He worked in the full knowledge that those doing the trapping and killing were not the real culprits. Arresting poor villagers or even exposing small-time local police involved in the illegal logging or wildlife trade would have no effect on those ultimately responsible.

In early 1990 Seub had a meeting with then agriculture minister Sanan Kachornprasart, in which he tried to explain the complexity and gravity of the problems he and his men faced in the sanctuary. It was a turning point. He was told by the minister, among other things, that he would just have to work harder. Seub said the encounter made clear the pervasive extent of corruption in the Royal Forestry Department, and the lack of support for his efforts. He came to the conclusion that just as the animals in the forest are victims of the wildlife trade, so he, his workers, and even colleagues in the Wildlife Conservation Division were pawns in a game so big that the real players could not even be named. After the meeting, he said: "I know now that we are fighting this alone."

Then in July, two of his rangers were shot at—one, at point-blank range. Though both men escaped serious injury, the shootings fundamentally altered the nature of this muted war. Seub felt he had been forced into a position where he might be sending the men he loved to their deaths. And for what? If a guard or ranger was killed, that death was on Seub's shoulders alone. The burden was all the heavier in the knowledge that they would have died, in part, out of allegiance to him. It was also a foregone conclusion that, in the event of a death, no one would take notice; forest protection would not be advanced. For the victim's bereaved families there would be little compensation or even comfort in knowing that their sons and husbands had not died in vain.

Suicide is the most private of acts. The decision, like the reasons, cannot be known to us. But from the outside Seub's death captured public attention because he defended aspects of the forest that Thai forestry had for so long undervalued. He was perceived as a rare and endangered creature: an honest forester struggling against the corruption within his own department.

PART 2

SCIENTIFIC FORESTRY ENTERS SIAM

The English Government will throw in our teeth [that] we do not know how to rule our own state, that we have resources but do not know how to preserve them properly... With [the Burmese] example before us, there is cause for fear... we should hasten to take measures for the reform of the system of forest protection... before it is beyond our power.

Letter from King Chulalongkorn to
Chao Muang Inthanon of Chiang Mai (1897)[1]

The essence of Siamese forestry—the structure of the Royal Forestry Department, the nature of the forestry system and laws, and the attitudes and educational background of the foresters themselves—has long mirrored that of a British colony. Decades after logging was banned, the Thai forestry establishment continues to view local people as the enemy. Unlike its neighbors, Thailand, then Siam, was never formally colonized during nineteenth-century European imperialism. While lands adjacent were turned into British and French provinces, the monarchs in Bangkok managed to maintain a degree of political autonomy and control over people and resources.[2] Yet the pattern of economic change that emerged in Siam was markedly similar to that of neighboring colonial states, and the Siamese elite pursued economic policies that resembled those of a European colonial regime.[3]

Certainly, in regards to the extraction of teak wood, a commodity of strategic importance to the colonial powers, the legal and bureaucratic institutions that emerged in Siam bear a striking resemblance to those of its neighboring countries. The fear of being colonized contributed in no small measure to Bangkok's efforts to centralize control over the northern forests and, ironically, to establish a quintessentially "colonial" institution for the control of timber resources. Regulations were put in place to secure exclusive state rights and manage timber extraction, and a force of men was trained to guard this treasure *for* the state *from* the people. The fact that interests represented by "the state" in Burma, for example, were those of a foreign, occupying power seems not to have been an important distinction for the nominally independent Siam.

By the second half of the nineteenth century, the British were badly in need of teak for ship and railroad building due to the lack of their own forests, having lost access to America's forests and having initially destroyed vast areas of timber resources in India and Burma through decades of so-called "laissez-faire" forestry.[4] "The waste in exploitation was appalling . . . the forests being regarded [by the government] as inexhaustible," remarked one forester.[5] British forests had in fact been liquidated by the eighteenth century and lacked a forestry tradition of their own. They were therefore forced to look elsewhere for this expertise. They found it in the area of Central Europe that is today Germany, where the science of forestry had been developed over the previous century. The goal of this new science was to create even-aged, single-species stands to secure the state a long-term supply of timber.

German forestry developed in a temperate ecosystem of comparatively low biological diversity, and was predicated on full state control over forest lands. This approach to forest management was superimposed on the ecologically and culturally diverse forests of Asia, including that of Siam, via the British imperial machine. As the British teak industry expanded, forestry administrations were established. The German forest botanist Dietrich Brandis took charge of forestry in Burma and then in India for the purpose of systematizing logging, mainly of teak. In the late 1800s, a temporary closure of Burmese forests to timber cutting caused the British logging firms to expand their operations across the border into Siam. Initially, concessions were dealt out haphazardly by the local princes, the *chao*, who controlled the forests of northern Siam. In response to complaints from the logging companies, a British forester named Herbert Slade took charge of forestry, seconded in 1896 from the Burma Forest Department. Slade set up a bureaucracy in Siam modeled on the Burmese system, and introduced rules and legislation identical to those used in British colonies ranging from India to Australia.

The roots of Thai forestry thinking then—like that of forestry in many tropical and temperate countries around the world—reflect the German forestry tradition as it was applied during the colonial era. Thai foresters still call the logging method that was used in Thailand the "Brandis System."

Slade viewed himself as a representative of "forestry as a Science,"[6] as one whose main purpose was to regulate the rate of timber exploitation so that the resource would last, essentially forever. As we know, the resource did not last. When Slade became conservator, forests in northern Siam were being plundered by European timber companies. And though he seems to have tried to reduce the level of cutting, over-cutting was prevalent after he left in 1901. Indeed, it continued for the rest of the century until commercial logging was banned entirely in 1989. The real lasting impact of the style of forestry introduced at the end of the nineteenth century was not systematic forestry use, but rather a political system where the Royal Forestry Department eventually claimed a monopoly over more than 40 percent of the country's land area and criminalized the millions of people who happened to be living there.

The Problem of Diversity: The German Forestry Model

> *[The tradition of German forestry science produced] the monocultural, even-age forests that eventually transformed the Normalwald from abstraction to reality. The German forest became an archetype for imposing on disorderly nature the neatly arranged constructs of science ... [transforming] a ragged patchwork into a neat chessboard.*
>
> Henry Lowood, *The Calculating Forester*[7]

From the mid-nineteenth century when scientific forestry was introduced into India and Burma, colonial foresters faced two basic problems in implementing their task of managing the logging of teak:

the presence of rural forest communities and the staggering species diversity of the forests where teak grew. The techniques that developed to overcome these two obstacles came to determine the very nature of forestry in the British colonies and, by extension, in Siam where a similar "colonial" structure was put in place half a century later.

Teak grew sporadically over vast areas, mixed together with a myriad of other species. Initially, the challenge was simply to find the trees and get them out of the forest. But eventually efforts turned to transforming the forest itself by ridding it of other less valuable species, and increasing the incidence of teak—in other words, increasing its commercial value by "minimizing nature's diversity and reconstructing the forest to make life easier for foresters."[8] Though this began with weed suppression and thinning, it would lead to replacement with single species, and eventually to the modern clone-based monoculture plantations. It is important to remember that removing what we today would call "biological diversity" was neither an unfortunate oversight, nor the result of the incorrect application of forest management rules, nor the fault of uncontrollable external forces. Destroying diversity in the pursuit of maximum timber production was, for colonial foresters, the principal challenge and the intended outcome of their efforts. It was both their means and their end.

The German model

Naval strength was the key to maintaining and expanding British imperial power. Typically, a warship lasted an average of twenty years and required about two thousand oaks of a size that corresponded to more than two hundred years in age.[9] As European oak supplies diminished, teak—which occurs naturally in the forests of India, Burma, and northern Thailand—became the prime raw material for building ships. Its strength, durability, and resistance to moisture also made teak ideal for building railway lines. An American consul-general based in Bangkok at the turn of the century called it "the most valuable lumber for shipbuilding in the world." He noted that teak wood "does not yield to the influences of moisture and drought; it is not liable to the attack of borers and other insects; it does not split or sprawl, and, while it is strong, durable wood, it is easy to work and very light in the water ... because of its peculiar qualities that resist the influences of iron ... there is no substitute for it yet discovered as the backing for armor plates in vessels of war."[10] He estimated, moreover, that one-fourth of the world's teak supply was standing in Siam's forests.

But logging by the colonial trading firms began on an *ad hoc* basis and wreaked havoc on tropical forests. It expanded in India and Burma through the first half of the nineteenth century with little heed to future supply. With few regulations and even less enforcement, the result was the liquidation of vast areas of the most readily accessible forests along coasts and rivers from Malabar to Pegu.[11] One part of the problem was that the British did not know forestry and had no system with which to control harvesting. The British Indian forester E. B. Stebbing wrote in *The Forests of India* that the administration "possessed no knowledge of tropical forestry, nor, indeed, of European Forestry, since British Forestry had almost ceased to be understood as a commercial enterprise in Great Britain."[12] But the government in London began to realize that the forests of its Asian colonies would not last forever, and something had to be done. It therefore came about that German foresters—precursors of the forestry aid consultants of our time—were hired by the British to put their colonial forestry resources in order.

Scientific forestry had emerged in Germany in the previous century, with the world's first university training program in forestry established in 1787 at Freiburg University.[13] The influence of German forestry thinking then spread quickly throughout Europe—first to France to the Écoles Forestières at Nancy where Herbert Slade and the influential American forester Gifford Pinchot were trained, then to Austria and Russia, as well as to the Nordic countries and Japan. The ideas spread to North America via Pinchot and the German, Bernard Fernow, and to the British colonies in Asia and Africa. The Dutch also hired Germans to systematize logging and teak plantations in Java. It is difficult to overstate the influence of German forestry thinking, which still today leaves indelible traces on forestry in Thailand and many other Asian countries. Writing in the 1930s, the German forester Fransk Heske described this global influence: "For all time, ... systematic forest management in Germany, [through] which the depleted, abused woods were transformed into well-managed forests with steadily increasing yields, will be a shining example for forestry in all the world."[14]

Originating in a country with about a dozen naturally occurring tree species, the purpose of German forestry was to establish even-aged, single-species stands to secure a long-term supply of timber for the state. The management system was based on an ideal abstraction of the natural—diverse, dynamic, and complex—forest. It was a biologically myopic view of the ecosystem that assigned value to a few tree species while attempting to eliminate others from a given area. According to Henry Lowood, three principles defined German forestry: minimum diversity, where the "arbitrary details" of nature should simply be cut out; the balance sheet, where mathematical calculations for estimating standing volume and future production potential were the forester's key tool; and sustained yield, the ability to "always deliver the greatest possible constant volume of wood."[15]

The German forester's forest was "populated not by the creations of undisciplined nature," but by the *Normalbaum*—the "normal" or standard tree.[16] This idealized being was the building block upon which a *Normalwald*, or "normal," ideal forest was created. To create a radically simplified forest, a kind of tunnel vision was required to "erase" from view the vast majority of grasses, flowers, ferns, mosses, shrubs, vines, lichens, and fungus as well as mammals, reptiles, birds, amphibians, and insects that make up a natural forest. German forestry science became a rigorous technical and commercial discipline centered on transforming the forest from a diverse natural system into a "one-commodity machine."[17]

Judged on its own criteria, scientific forestry in Germany was by the mid-nineteenth century a ringing success. Growing periods ranged from 100 to 150 years, and the inherent weaknesses with monoculture planting did not become apparent until well into the second rotation. From the years after the Seven Years' War (1756–63) when German forests were, famously, "so ruined that ... hardly a single bird can fly from tree to tree," to the turn of the last century, German foresters rebuilt the forests according to their own design.[18] They did so with the non-native Norway spruce and Scotch pine, two species that were to German foresters what *Eucalyptus camaldulensis* is for Thai forestry today. The moist, temperate climate lent itself to intensive management. Vast areas were planted, predominantly with spruce, which occurs naturally only in the mountains, but proved easy to plant in degraded lowland areas. Two centuries of practice resulted in a "restructuring" of German forests. While conifers used to be rare, they now make up two-thirds of the wooded area. Monocultures prevail with two-thirds of the softwood area stocked with only one species. And oak and beech, formerly the dominant species, now account for less than 5 percent of all trees less than 40 years old.[19] "Spruce used to exist only in the mountains. Now it's everywhere. People think it's natural but, in fact, it is a completely artificial situation," says German Forestry Professor Siegfried Lewark.[20]

This approach to forestry was grounded in a curriculum that traditionally focused on two distinct areas: production (including harvesting, reproduction, silviculture, and planning) and biology (that is, botany and zoology). The equal emphasis on what today would be called "forest ecology" appears at first difficult to reconcile with the timber bias of the profession. The reason for this was that foresters needed to understand how the natural system functioned in order to control insects and fungus that could damage timber, as well as how wood formed at the microscopic level.[21]

Brandis and the colonial foresters

The key figure in initiating the transfer of German forestry to Asia, and indeed to North America, was a man whose studies concentrated on the "biological" aspects of forestry. Dietrich Brandis (later knighted by Queen Victoria) was the German botanist who took charge of the Burmese Forest Service in 1856 and then became India's first Inspector-General of Forests in 1864, a post he held for nineteen years. He went about his work "at the cost of untiring labor and severe hardship with utter disregard to personal comfort and convenience," thereby eliciting "the highest admiration" from his superiors and colleagues.[22] Traveling widely throughout the sub-continent, he formulated a system of forest management and valuation whose ideas are still influential today.

For Brandis, emulating the forests of Europe was not some abstract idea. He initiated the practice of bringing forestry officers from Asia to visit German and French forests—"some of the best-managed forests on earth"[23]—as a part of their training.[24] Brandis was hardly ignorant of the rich biological legacy of Asia's teak-bearing forests. Among many other texts, he authored a handsome botanical reference book of Indian trees, complete with detailed descriptions and lovely paintings of seeds, seedlings, and leaves of hundreds of tree species. Though we know that German foresters learned about the forest's complexity in order to master it, it still challenges the imagination to understand how, with his considerable knowledge, Brandis and generations of his successors could put such heroic efforts into diluting this magnificent diversity and replacing it with just one species.

The prerequisite for accomplishing this task was securing state control over the land. It is no coincidence then that the first tenet of colonial forestry, outlined by Brandis in October 1856, was that forests must be the property of the government.[25] His next three tenets consolidated state power over land, at least on paper, outlining administrative structure, assigning exclusive rights over forest use to the superintendent of forestry, and explicitly criminalizing all other forest users.

While Brandis is seen as the father of systematic forestry in the colonies, his successor in India, another German, Wilhelm Schlich, is credited as being the founder of imperial forestry education.[26] In 1885 Schlich took leave from his position as inspector-general of forests in India to set up the empire's first forestry school at Cooper's Hill, the Imperial Forest School, from which officers were recruited. In 1878 a school for training "subordinate staff" was opened at Dehra Dun in northern India, to which Siamese, Burmese, Malayan, and other "natives" of the colonies were sent to study.[27] Schlich, who was also knighted, founded the *Indian Forester* journal in 1875, which continues to be published out of Dehra Dun. He also wrote the four-volume tome,

Schlich's Manual of Forestry, which was for decades *the* authoritative textbook on forestry worldwide.

The hair-raising language of colonial foresters reflected their preoccupation with commercial timber production to the exclusion of other elements of the forest. The fourth volume of *Schlich's Manual*, for example, entitled *Forest Protection*, includes the following chapter headings: "Protection of the forest against animals," "Protection of the forest against deer and wild pigs," "Protection of the forest against rodents," "Protection of the forest against birds."[28] The notion of "forest" was clearly interchangeable with "timber resources." It followed that "principle," "valuable," and "desirable" species were timber trees, judged solely by their market price and bearing no relation to their ecological or other aspects. Non-timber products were, at best, secondary. Areas that might for local people represent an important supply of medicinal or food plants could readily be dismissed as "inferior" or "waste land" by foresters.[29] If the goal of the forester was to create a *Normalwald*, it followed that the diverse forest that had not benefited from a forester's intervention was "abnormal" and chaotic. Schlich's successor Robert Scott Troup likened the forester's mission to the burden of missionaries bringing the light of civilization to unruly savages: "The attainment of the normal forest from the abnormal condition of our existing natural forests as a rule involves a certain temporary sacrifice ... in introducing order out of chaos."[30]

Bernard Fernow, another influential German forester who initiated forestry education in North America,[31] made a distinction between "tree weeds and runts," on the one hand, and "desirable and useful" species on the other.[32] Instead of modifying their forestry knowledge to fit the vastly different conditions of the New World, early American foresters "set out to mold the forests to fit their theories."[33] US forestry thinking tended thus to be dichotomous, dividing the forest into two categories: timber and "ROTT" (resources other than timber).[34]

In the Asian colonial context, resources other than timber included forest dwellers. Foresters were confounded by people who were "accustomed, without let or hindrance, to get what [they] wanted from the forest."[35] In their reports, they frequently made the point that because of the length of time it took for trees to mature, the state was better placed to take care of forests than private persons. It was something of a tautology for foresters that they possessed the most knowledge about forests. Yet where local people might see fruits, vegetables, dyes, oils, balms, medicines, fuel wood, textile fibers, construction material, or the family graveyard, foresters could just see weeds—the "arbitrary details" of nature. Arguing specifically against community ownership of forests, Wilhelm Schlich pointed out in *Forest Policy in the British Empire*, the first volume of his *Manual*, that "the personal interests of the members of the [village] are likely to injure the sustained yield of the forests; the poorer members especially urge more extended utilization than the forests can stand permanently."[36]

Uncontrolled commercial logging was of course a problem. But *taungya*, whose original meaning in the Burmese language refers to rotational shifting cultivation involving controlled burning and fallow periods, was systematically singled out by foresters.[37] They knew that no forest regions where the valuable timber grew were free of human activity, but villagers were clearly a much easier target than influential European merchants. In a typical description, Stebbing, who acknowledged the devastation caused by logging, referred nevertheless to shifting cultivation as a "pernicious system" in force for centuries, which was "the most wasteful of all methods of forest utilization ... and thousands of square miles of the valuable forests of the country had been laid waste by the system, the formerly fine timber forests being replaced by worthless scrub."[38]

Brandis's later views on the matter are in stark contrast with that of his colleagues. While he shared his profession's faith in scientific forestry, his awareness of rural communities' capacity to manage forests was quite unique among foresters, then as now. Reflecting on decades of work in Asia, Brandis wrote appreciatively about the widespread network of sacred groves in India and of the "traditional system of forest preservation" that were part of an "indigenous Indian forestry."[39] He envisaged a parallel system of management that combined centralized, state-controlled forestry with a network of village forests that would cover a significant, though smaller area. The latter would be controlled by local communities and could provide them with firewood, thatch, wood, bamboo, and grasses for tool making and manure, as well as land for grazing. The British government did not share Brandis's faith in local competence and was, moreover, reluctant to relinquish control over valuable timber areas. His proposals for restricting the state's takeover of forest land were summarily rejected.

Forest law in Burma and India, as Siam, defined forests as state land. Foresters were therefore in their full right to forbid villagers from destroying state property. But this was practically impossible because foresters lacked the manpower and budgets to patrol such immense areas. Moreover, local people could cause all manner of damage if they were pushed too far. Brandis was opposed to the complete prohibition of shifting cultivation, fires, and collection of "minor forest produce." He warned prophetically that "in the long run Forestry cannot succeed unless the people who live in and near the forest are for it and not against it."[40] He wrote that it was "quite out of the question to prohibit *toungya* [sic] cultivation without special permission over this extensive area, nor was this contemplated." Such measures would "produce a most harassing and vexatious interference with the inhabitants of the forests without any corresponding advantages."[41] With uncanny foresight he even warned in the 1870s that "the annoyance to the inhabitants by the maintenance of restrictions over the whole area of large forest tracts will be permanent, and will increase with the growth of population."[42] These warnings went unheeded and forestry's tasks came to focus on extending state control, transforming the physical landscape, and limiting customary rights as far as politically feasible without invoking serious rebellion.

Dissenting voices

Forestry would continue in this direction for more than a century. It would, however, be wrong to suggest that there was no debate or dissension among European foresters working in Asia. Even in the heyday of scientific forestry, the absurdity of trying to recreate the biological conditions of Germany or France in Burma or Siam did not go unnoticed. Critical questions were raised about the basic premises of scientific forestry—about the risks of monoculture plantations, the possible benefits of fire, the real impact of shifting cultivation on the teak-bearing forests and, in general, the danger of applying uncritically in Asia assumptions that seemed valid for European forests.

In India, Dietrich Brandis viewed scientific forestry as a "plant of foreign origin" that needed to be naturalized in order to thrive.[43] This most famous advocate of the system believed that the system could be adapted in order to function in a new environment. But it was not obvious to everyone by the turn of the nineteenth century that the lessons of Europe's forests were, in fact, applicable or even relevant to the tropics. One British forester commented that while forestry in France and Germany is "truly scientific . . . conditions [in Asia] are so totally different that we cannot utilize the results obtained in Europe."[44]

A critical review in *Indian Forester* of the 700-page fourth volume of *Schlich's Manual of Forestry*, entitled *Forest Protection*,[45] argued along similar lines. The 1907 article notes that in order to make it more relevant to British Indian foresters, the English translator omitted material that was "only of local interest." Still, one third of the book is devoted to European insects. The reviewer quite sensibly questions the relevance of this information for foresters working in completely different ecosystems: "The dangers that threaten the well-being of a forest on one side of the globe will very likely be different from those which have to be guarded against in forests on the other side . . . It seemed hardly worthwhile for a student to work through over 200 pages of European insects."[46] The book also contains fifty-six pages on European fungi.

As for the risks of monoculture tree plantations, an editorial in *Indian Forester*, entitled "The danger of formation of pure forests in India," warns that foresters might be pushing the goal of monoculture too far. It notes that "we are all of course aware that the text-books on the subject are unanimous on the disadvantages of pure forests" due to the dangers posed by insects and fungi. "It is very often the case in India that there is only one really principal species in a given forest crop. [There is] a marked tendency in . . . improvement fellings to go too far in the direction of the establishment of pure crop of the principal species."[47]

Since colonial times, it has been the self-evident truth of the forestry profession that local people destroy forests. It was, however, occasionally suggested that the application of villagers' local knowledge might produce better results—for timber production—than regulation based on scientific forestry principles. Decades after Brandis argued for a parallel village-based forestry, G. R. Long pointed out that in forest areas along Siam's northwestern border area where there was "very little forest law . . . [and] people cut where it seemed good to them," the teak trees growing in farmers' fallow clearings were, in fact, of superior quality to those growing in "pure plantations."[48] He was critical of the expensive and time-consuming policy of confining shifting cultivation. In the Salween hill tracts, where Karen were permitted to practice swidden agriculture, "their abandoned clearings soon filled up with teak seedlings . . . In the Southern Shan State matters were similar . . . Consequently, you will now find many a three-acre patch of teak on the Salween simulating an even-aged, fairly stocked, plantation of from 10 to 50 years of age."

Long concluded that if *taungya* were permitted more widely, rather than being systematically discouraged, "the forest obtained would be of an improved natural type—which is more than one can claim for pure plantations of non-gregarious species."[49]

Another British Burmese forester went even farther than Long in questioning the conventional wisdom. Herbert Slade, who would bring scientific forestry to Siam, published an article in *Indian Forester*, entitled "Too much fire-protection in Burma," probably written during his last year in Burma.[50]

Fire suppression was a central tenet of colonial forestry, and the British government aimed to extend fire protection to all teak forests in the empire. Slade noted that the common belief was that annual fires were caused by human carelessness and that by protecting forests from fire, foresters would "[counteract] the evil" and restore forests to their original condition. "I venture to disagree with this theory *in toto*," he wrote.

Slade advocated fire suppression solely in young teak plantations. His main premise was that general fire suppression during other stages of teak's growth was a waste of time, money, and already limited manpower. It did not improve timber stocks, but rather degraded them in the long run. He observed that fire enhanced teak growth in several ways: teak seeds lying on the ground became charred, but undamaged inside; roots of saplings were undamaged; young saplings were destroyed completely, but grew the following year with increasing vigor until they finally survived; standing teak trees were unaffected; and all other vegetation was seriously damaged, which was another advantage for teak, a light-demanding species. The logical conclusion was not merely that annual fires were conducive to teak growth, but also that the natural reproduction of teak over large areas might be impossible without the assistance of fire.

Annual fires and teak, he wrote, had coexisted for so long that it was mere speculation to suppose that teak even existed in the "remote ages" before there were people living in Burma. "Annual fires... are now so constant and regular as to have become natural to the teak... May not the thickened seed and the corky bark, both alike unaffected by the ordinary ground fire, have been gradually developed for this purpose?" He noted, moreover, that fires in Burma were typically ground fires "slowly but surely advancing and consuming the dry leaves which cover the ground to a depth of a few inches or less. As a rule the flames are not more than a foot or so high and a pony will step over them."

His assessment of the way foresters handled fire was uncompromising: "We find the teak growing under certain conditions, we are convinced that it has done so for ages, and yet we seek to change those conditions without any proof that we are not thereby actually damaging the object of our care."

The article provoked a "mass of literature" in the journal and launched a debate that would continue well into the next century.[51] What may have irked Slade's superiors more than the actual proposition that fire could be beneficial was the cockiness with which he presented his ideas. Throughout the article, he accused his superiors of relying on unproven opinions and unsubstantiated theories, and of lacking evidence, proof, or statistics for the widely held belief that forests must always be protected from fire. Slade wrote: "a mere expression of opinion is not sufficient."[52]

To distance themselves from Slade's conclusions, which did not apply "anywhere this side of the Bay," i.e., in India, the editors of *Indian Forester* printed this disclaimer after the article: "We are bound to say that we disagree with our author in objecting to opinions, and requiring statistical proof. In Forestry, in our view, *it is the mature opinion of experienced professional men that is more valuable than statistics* (emphasis added)."[53]

The article appeared in *Indian Forester* in May 1896, just one month after Slade had completed his tour of northern Siam and delivered his first report to Prince Damrong.

Herbert Slade's Legacy

> *The Conservator is not only an employee of the Siamese government but also the representative in Siam of forestry as a Science... He alone in Siam knows what measures should be adopted in the interests of the forests... having seen their effect in other countries.*
>
> Herbert Slade, Siam's first conservator of forests (1901)[54]

Herbert Slade appears neither to have been a rogue nor a spy. Had he been corrupt, Slade might have made a neat personal sum by forging alliances with the six European timber companies that were logging teak in the northern forests or with the chieftains, or *chao*, who owned those forests. I have seen no evidence that he did this. Trusted as he was by the Bangkok administration, he could presumably also have worked undercover to extend British colonial ambitions into Siam, using the pretext of giving advice on forestry. If he tried, he did not succeed. Moreover, judging by his critical 1896 article on Burma's fire policy and by his final report to Prince Damrong in 1901, Slade was not afraid to speak his mind either to his British superiors or to his Siamese employer.

Though Slade spent less than six years in the kingdom, he managed to lay the legal and bureaucratic foundations of Thai forestry that would last a century. Slade was personally insulted by the suggestion that he favored British firms over other interests. From his reports, he appears to have gone to some lengths to bring the timber companies' activities under control. He navigated a precarious path among the competing interests and, in the process, he played no small part in the annexation of northern Thailand—a "silent revolution"—that resulted in the consolidation of Siamese territory.[55] Through a combination of diplomacy and bribes, the northern princes and chieftains were persuaded to give up their control over forest lands, and the Bangkok administration achieved a monopoly over teak lands and the royalties that came with it. One observer wrote that "Slade may be said ... to have played an important part in consolidating the Siamese Kingdom and to have been of great assistance to the Government in this regard."[56]

Slade introduced a regulatory system that attempted to slow the rapacious logging, while at the same time ensuring the timber companies a steady supply of Siamese teak. King Chulalongkorn was so satisfied with Slade's work that, for his six years of service, he conferred upon the British forester the Companion of the III Class of the Order of the White Elephant of Siam.[57]

Still, the presence of Slade and his two British successors likely contributed to the domination of the Siamese teak industry by European, mostly British, logging companies. Moreover, after almost three decades of British forestry leadership, the over-cutting continued. And by the late 1920s, though Siam's forest lands were officially under state control, the Royal Forestry Department held logging concessions on only 1 percent of the teak forests, while European firms controlled 85 percent.[58] The British domination of the Thai teak trade lasted until 1955 when most foreign timber licenses were revoked. By then, however, the largest trees had been logged out and what was left in Thailand's forests were teak trees of a size class that should, according to earlier rules of scientific forestry, not have been harvested for many more decades.

Before Slade

The temporary closing of the teak forests of Upper Burma at the conclusion of the Third Anglo-Burmese War of 1885 brought the major European timber companies into northern Siam.[59] When Herbert Slade arrived in Siam, three British companies controlled almost all forest leases on the tributaries of the Chao Phraya River: Borneo Company, Bombay Burmah Trading Corporation Ltd., and Siam Forest (later renamed Anglo-Thai Company).[60] Within the next decade, three more European firms would be granted timber concessions: the British firm, Louis Leonowens Company Ltd., the Danish East Asiatic Company, and the French East Asiatic Company (see inset, p. 63).

Pointon's history of the largest of these six firms, the Bombay Burmah Trading Corporation, describes the situation before 1896: "Northern Siam was something of a Tom Tiddler's ground... There was no Forest Department. The chiefs were still supreme. Their forests, whence timber was being taken in quantities wholly unregulated, were being most wastefully and haphazardly denuded. Whoever had interest with one of the chiefs... needed no qualification beyond the presentation of a suitable gift to secure a permit for felling teak."[61] The felling of teak trees was "entirely indiscriminate," and "every trunk which could conceivably yield a plank or even a butt-end was hewn down and cast into the streams."[62]

By the 1890s, the British timber firms were acting as if they owned the place, and King Chulalongkorn feared outright annexation. A secret report by the British Consul in Chiang Mai from 1902 suggests that the king had serious grounds for concern:

> The extinction of Siam as an independent Power is merely a question of time... Situated as Siam is—a weak, scarcely civilized State, hemmed in between two strong colonizing Powers—its fate is inevitable... The march of history cannot be arrested and unless European influence in

the Far East is to entirely disappear, Siam must sooner or later succumb to a European power.[63]

There were already several examples of the control over teak resources to supply the British navy and prevent France from gaining access, leading to British colonial expansion. In the late 1700s, the British had gone to war to acquire the teak forests of Malabar from the Dutch and the French. Wars in Burma secured British control of teak in Tenasserim (1826) and in Pegu (1852); the annexation of Upper Burma in 1885 solidified British control over Burma's teak forests; and the annexation of the east side of the Salween in 1891 extended this control into territory that had, as late as 1884, been viewed by the British as part of the principality of Chiang Mai.[64]

The growing concentration of power over teak logging in the hands of British companies' meant also that revenues were not reaching Bangkok.[65] The central government had introduced measures over two decades to increase its share of timber income. But in 1895, Borneo Company and Bombay Burmah paid royalties on as little as 20 percent of the twenty-five thousand teak logs that the companies sent down the Chao Phraya River.[66] Bombay Burmah was positioning itself to be able to "absolutely dictate their own terms" to the government.[67] The *chao*, meanwhile, took full advantage of the situation, playing one European firm against another to obtain the highest possible leases and frequently double-leasing forest areas.

The regulations governing teak logging in northern Siam at the end of the nineteenth century were both loose and loosely enforced. The 1883 treaty between Great Britain and Siam had defined rules for the cutting and girdling of timber by British firms.[68] But it included only a "wistful note" about preventing forest owners from making agreements with more than one party for the same forest. The British Advisor to the Siamese Ministry of Commerce Reginald Le May described the treaty as "a definite attempt to bring [the] local Chiefs to heel, and to [give] . . . the British companies and individuals engaged in the teak trade a measure of fair play in their dealings with them."[69] Significantly, the treaty also prohibited British subjects from engaging in forestry without the Siamese government's permission.[70] This contributed to the gradual exclusion of the smaller-scale Shan and Chinese timber merchants who had hitherto controlled the trade in Siamese teak.

Another reason that the European timber companies were able to dominate logging in Siam's northern forests is simply the amount of time and money required to get access to the trees and bring them out of the forest. Foresters call teak a "non-gregarious" species because the trees do not occur naturally in concentrated clumps (as pine does), but appear sporadically over vast areas.[71] Early colonial foresters in Burma, for example, estimated that some five hundred thousand marketable trees were scattered over 7,000 square miles.[72] In northern Siam, the Royal Forestry Department had over a million teak trees to girdle that were spread throughout the region.[73] Since they did not grow in neat rows, finding the trees and assessing the total extent of the resource were no easy tasks.

The process of teak extraction also involved a long series of steps. After identifying the trees, they were girdled (a groove was cut around the trunk), which killed the trees and allowed them to dry out, thus also making them lighter.[74] They were left standing for up to two years and were then felled and dragged by elephant to the banks of the nearest stream. When the next rainy season came, if the waters rose high enough, they could be floated to one of the northern rivers (Ping, Wang, Yom, or Nan), bound together in rafts and sent on to the Chao Phraya and then downstream to Bangkok. If these trees were cut in western watersheds closer to the Burmese border, logs would be floated down the Salween and then onto Moulmein. From forests to the east, logs would be floated down the Mekong into the French territories. The average time between girdling and arrival in Bangkok was reported to be about five years, but a few dry seasons could slow the process by two or three years. In the worst case, the whole process could take as long as fifteen years.[75] It is therefore not difficult to see why huge financial backing was necessary for logging to be profitable.

The Founding of RFD

Something had to be done to put Siamese forestry in order, and there was little doubt that a Western forester would have to be hired. But the Bangkok government was caught between two colonial menaces. The Siamese-French naval confrontation of 1893–94 had only recently subsided, and there was concern about offending the French by hiring a Brit.[76] The British, moreover, were worried about the implications for their timber companies of hiring a German, though this was perhaps the natural choice given the role of German foresters in India and Burma. King Chulalongkorn's first choice was thus Jørgen Castenskiold from the Kingdom of Denmark, a small country without obvious colonial designs.[77] Castenskiold was hired in 1895 to survey Siamese teak resources, but died later that year. It seems plausible to suppose that had the Dane survived, he would have been appointed Siam's first conservator of forests. As for the choice of Herbert Slade, it has been suggested that Bangkok saw this British national, who had received his forestry training from the Écoles Forestières in Nancy, France, as a man who could assuage the delicate sentiments of both the British and the French.[78]

Slade was seconded to Siam from the Burma Forest Department in 1895, where he was then deputy conservator of forests in charge of the Tharawaddy Division. He took up his new position as an employee of the Siamese government on January 3, 1896, and traveled, as Castenskiold had started to do, in northern Siam to survey the teak resources for over four months.[79] He then submitted a report on the state of Siam's forests to Prince Damrong, minister of the interior, in which he proposed a new set of rules to regulate forest use. These, he admitted, would draw heavily on existing forestry laws that were already in use elsewhere:

> This forest law [is] based upon the forest law in northern Burma. I need not apologize for this as this law is accepted by all foresters as being one that has been used successfully to deal with complex forestry problems. And just last year, the Director of India's Forest Service asked that northern Burma's forest law be used in Australia as well.

If it can be applied there, it can certainly work well in Siam which is very similar to Burma.[80]

Faithful to the tradition of Dietrich Brandis, Slade stated in his report that the first issue that needed to be resolved as quickly as possible was the problem of forest ownership. He wrote: "In Burma, all the forests belong to the state, but in the Kingdom of Siam, it is accepted that forests are owned by particular persons. They are called 'forest owners.' These forest owners have the power to forbid logging on their land, and to demand compensation when they permit timber extraction such that they receive the highest benefit."[81]

Royalties, moreover, were not reaching Bangkok. Slade estimated that in 1896, of the 80,549 large-sized teak logs floated downriver to Chainat from the north only 4,776 were being stamped as having had royalty paid on them.[82] The central government was thus receiving only one-sixteenth of its rightful share of the income from teak. "If a Forest Department is established as I propose," Slade wrote to Prince Damrong, "the benefit to the country [read: the Bangkok government] would increase considerably due to the collection of royalties and tax."

Slade's proposal fit well with Bangkok's interests. King Chulalongkorn needed to boost the treasury's income to finance widespread reforms in the country. But neither local companies nor the government had sufficient capital to log the forests themselves. An earlier attempt to hire the American doctor-cum-logger, Marion Alonzo Cheek, to operate a timber license on the government's behalf had ended in miserable failure and a protracted law suit with the US government (see inset, p. 63). Northern princes were playing one foreign logging company against another to gain the highest possible concession fees, and little of this money found its way into Bangkok's coffers. Equally important for Bangkok was reducing the ever-present threat of annexation by consolidating and formalizing control over territory. While the colonizers rationalized their conquests by arguing that they were bringing progress to less privileged peoples, the Siamese needed to preempt European accusations of backwardness, and show that they could manage their affairs in a modern, "civilized" manner.

Slade completed his tour in April. The Royal Forestry Department was established on September 18, 1896, in the Ministry of the Interior, and Herbert Slade was appointed as conservator one month later.

Slade felt that Siam's forests were being badly overworked. The timber production capacity of the forests was being "greatly exceeded" and he feared that unless the fellings were reduced "the existence of the forests would be jeopardized."[83] In early 1897, Slade set about to reduce the logging and girdling in areas where only a few trees were left.

One of his first moves was to declare null and void all leases made after February 1897, and to convince the main firms to accept the new government leases.[84] According to the terms of the new leases, royalties on logs were raised, the minimum girth for girdling was set at 2.1 meters, the felling cycle was increased from six to twelve years divided into two six-year halves, and penalties for infractions were introduced.[85] The Forest Preservation Act and the Teak Preservation Act were passed, and in 1898 a law prohibiting the illegal sale of timber was enacted.[86]

At least as important as the change in the content of leases was Slade's success in removing the power to grant leases from the *chao* and placing it exclusively in the hands of the central government. The British advisor Le May observed that "in his work in organizing the department Slade naturally met with a good deal of opposition from the local northern chiefs on whose preserves he had naturally to encroach to a large extent; but in the end, after a hard fight he won his battle and this victory weakened the position of the chiefs who never regained their former prestige."[87]

Another key element that Slade put in motion to ensure the continuity of colonial forestry practices and structures in Siam was the education of Siamese men at Dehra Dun in northern India, and later at the forestry school in Burma.[88] In 1900 four Siamese men were sent to Dehra Dun to participate in a three-year forestry training course with the ultimate aim of replacing all Europeans with locals once a sufficient number had received the education.[89] (All four eventually became directors general of forestry in Siam.[90]) During Slade's time, the department's staff consisted of sixteen Europeans (including W. F. L. Tottenham, another British Burmese forestry officer who would replace Slade in 1901) and nine Siamese. Altogether thirty-five Siamese men would be sent off for training before Phraya Daruphanphitak became the first Siamese forestry chief on April 1, 1924.[91]

All in all, Slade closed off half the area of forests originally being worked, and tripled Bangkok's teak revenues from 325,000 rupiahs in 1897–98 to over 1 million rupiahs in 1900–1.[92] He also prohibited the girdling and felling of undersized teak trees thus putting an end to the "enormous" trade in poles and saplings that threatened to exterminate teak over large areas.[93]

Slade's disappointments

For all his apparent accomplishments, however, Slade was despondent about how little he was able to achieve in Siam. Though he admitted that some progress had been made, it was "nothing in comparison with what should have been accomplished in the time."[94] Slade complained of being treated like a "junior clerk," and of spending hours each week writing pointless memos to Prince Damrong on "every tiny matter." He also insisted in his final report written in 1901 that he could have taken a much tougher line with the timber companies, especially Bombay Burmah Trading Corporation, had Prince Damrong been more supportive.

When Slade began negotiations with Bombay Burmah in July 1901, the company was active in Chao Phraya basin forests, and held a practical monopoly on Salween timber near the Burmese border. In Mae Hong Son in the northwest, the firm was "spending money recklessly" to gain a footing in the area.[95] And because of indiscriminate girdling, some twenty thousand teak trees stood dead in the forests of Mae Lan, which flows to the Salween, causing the loss of seed-bearing qualities and eliminating the possibility for natural reproduction. Slade's special deputy in Chiang Mai, J. G. F. Marshall, reported that "in the practically absolute dearth of seed-bearers there is little or no recent reproduction and very few saplings."[96]

In Slade's judgment, the company's position was weakened by its breaches of several leases "for which very heavy damages could have been obtained in the Law Courts."[97] Slade hoped he could use this as leverage to convince Bombay Burmah and the other firms to accept the new, stricter leases without resorting to the courts. In September 1901 Borneo Company agreed to pay higher royalties on logs and to halt all girdling, and Bombay Burmah followed suit in the Chao Phraya Basin forests. But in the Salween, according to Slade's report, Prince Damrong's vice minister went behind the conservator's back and promised Bombay Burmah that it would not be sued for its dealings in the Salween forests. This "secret understanding" between the Interior Ministry and the Bombay Burmah Trading Corporation seriously undermined Slade's authority and forced him to "reduce his terms considerably."

The result, according to Slade, was that huge areas of teak-bearing forest were needlessly destroyed by timber merchants who girdled as many trees as they could as a way of increasing their chances of having leases extended. The prince had the power to stop this, but failed to set the necessary policy guidelines.[98] In his own assessment, Slade had achieved "a good settlement both for the forests and for the Revenue." But he fell far short of his own expectations.

Slade spent most of his five and a half years in Siam overseeing logging operations, surveying, marking timber, and settling disputes between the companies and the northern princes. But he regretted that he had no time to "improve" the forest, and left it to his successors to mark off areas for timber concessions, make up working plans, establish plantations and, most important of all, to keep villagers out of the forest.[99]

Over the course of the century, the central government duly took measures to secure its control over forest land, timber production, and income. The duration of leases was increased from twelve to thirty years starting in 1908; thirty years remained the duration of the concession cycle until logging was banned in 1989. Initially, the northern princes were given a share of timber revenues, but by 1918 they were receiving a fixed annual payment that would not be inherited by their children, and lower-ranking nobility received nothing at all.

After the 1932 revolution that ended the absolute monarchy, Chiang Mai and the other northern provinces received no more special treatment.[100] State control over forest land was consolidated in the 1941 Forest Act, which defined forest in political rather than biological terms, as land that no one except the state has the right to occupy or use. Similarly, the colonial designation of forest reserves—areas reserved for timber harvesting—was eventually formalized in the National Forest Reserves Act of 1964. By the end of the twentieth century, 230,000 sq km, or 44 percent of the country's surface area, was categorized as forest reserve land under the exclusive jurisdiction of the Royal Forestry Department.[101]

As for timber income, it did not escape the notice of the government that in spite of the nationalization of forest lands, profits from logging were flowing into foreign hands. As early as 1912, a logging division was established in the department to generate revenue for the state, with its first operations in Mae Haek (Phrae) and Mae Chaem (Chiang Mai).[102] Though Thai leadership of the department began in 1924, British firms would continue to dominate logging in Thailand for three more decades. By 1930 the Forestry Department still had concessions on only 1 percent of the teak forests, while local lessees controlled 14 percent and European firms accounted for 85 percent.[103] After 1932, there were plans to nationalize the teak business, but these were stalled by the worldwide economic crisis and by the Second World War. Though British timber merchants fled to Burma when Thailand allied itself with the Japanese, the concessions were returned after the war as reparations and as compensation for damaged property.[104] Many years after the war had ended Boonsong Lekagul, the famous Thai conservationist, offered this telling description of the attitude of the British timber merchants who returned as victors to Thailand in 1945 to claim what they had left behind: "Where is my property? Give it all back. Not one log must go missing. Go and find every last one. If teak trees that were standing in my concessions when I left have been cut, you must compensate me from other concessions. Whatever cannot be replaced with timber must be paid for with money—I'll take not one penny less than I am owed."[105]

A reorganization of the forestry bureaucracy in 1947 resulted in the logging division being turned into the Forestry Industry Organization (FIO), whose purpose was to regain some of the compensation payments.[106] FIO was completely separated from the Forestry Department in 1956. When all foreign logging licenses ended in 1960, FIO became the main teak concession holder with 29.5 of the 39 concessions, and remained so for three decades—until the logging ban.[107]

Throughout the period, the over-cutting did not let up. During his time as conservator, Herbert Slade did not manage to limit logging to a "sustainable" level, and when the British companies left in the late 1950s, the relentless depletion of what remained was simply continued by local Thai and Chinese entrepreneurs. The size of timber coming out of Thai forests dropped steadily and forest cover plummeted from 53 percent in 1961 to 28 percent in 1989.[108] Half the concessions were closed in 1979 for lack of harvestable timber, and the 1989 cancellation of concessions formally ended the logging era.

Herbert Slade's legacy is a forestry bureaucracy whose structure, purpose, and legal bases resemble that of other British colonies in Asia. He also set in motion a process of knowledge transfer that ensured that the nineteenth-century European forestry tradition would be carried on by Thai foresters. In that tradition, the mandate of forestry was to secure state control over forest lands and to regulate the extraction of timber from the forests so that it could, in principle, continue indefinitely, and even increase over time. But Slade failed to achieve this, and left Siam in 1901 with grim parting words in a report to Prince Damrong: "When the forest history of Siam comes to be written, it is feared that [1901] will be handed down as the year in which the ruthless destruction of hundreds of square miles of teak forest was allowed."[109]

Six European firms that logged teak in Siam at the turn of the twentieth century

The Borneo Company, incorporated in England, traded in Siam's three main export commodities—teak, rubber, and rice.[110] The firm started logging teak in Siam after the Bowring Treaty of 1855 but suffered initial heavy losses due in part to the lack of clarity over logging regulations. Borneo returned to Siam in 1884 after the new treaty was signed, stationed an agent permanently in Chiang Mai, and this time had more success.[111] Thirty years later, Borneo had stations in Chiang Mai, Lampang, Ban Na above Raheng, and Muang Fang, where the firm constructed a light railway for the extraction of the timber.[112]

Part of the secret to Borneo's success was the fact that the company hired an American, Marion Alonzo Cheek, a missionary doctor who quit his evangelizing to enter the lucrative teak industry. Referred to locally as "the doctor who put down his needles and picked up a saw," Cheek was well known in the north where old people reportedly still remember him. He had many wives, and there are even folk poems (*bot klon phuen ban*) written about him.[113] He also had very good relationships with the *chao*, which he used to great advantage in negotiating leases.[114] Cheek entered into an agreement with Prince Naradhip, in which he was loaned money from Bangkok's treasury in 1889 and in 1890 to conduct teak logging on behalf of the Siamese government. The deal was that profits were to be divided one-third to Prince Naradhip, two-thirds to Dr. Cheek, with the exception of forty-four hundred logs that were to be delivered to the Borneo Company.[115] But bad weather and loose handling of finances—depending on whose version you believe—contributed to a collapse in the agreement. The government seized Cheek's property and declared him persona non grata. Cheek sued and the case between Siam and the United States government dragged on, even after his death in 1895. The Danish captain Jørgen Castenskiold, who had worked for the Danish East Asiatic Company, handled the Cheek legal case on behalf of the Siamese government before embarking on his 1895 survey of timber resources in the north.

The Cheek affair likely discouraged the Siamese administration from undertaking more of its own operations and may have contributed to the concentration of European control over teak logging. But Western domination of the industry was certainly also due to the government's lack of capital resources and trained manpower.[116]

Cheek's successor at the Borneo Company was none other than Louis Leonowens, the son of Anna Leonowens (of *The King and I* fame), the British governess who taught the many children of King Mongkut. Young Louis grew up in the palace in the company of King Chulalongkorn, and their subsequent friendship resulted, according to one source, in a gift of important teak leases.[117] He became a cavalry officer in the Siamese army, and rose to the rank of major before resigning in 1884. He was then hired as an agent by the Borneo Company and obtained numerous leases for the company "by living almost like a native, and by adopting to a great extent the native ways of doing business and systems of intrigue."[118] Leonowens became "a figure of considerable consequence and prominence in the north, enjoying a perhaps dubious prestige." He traveled around with a large retinue of elephants, ponies, and uniformed men. Lao maidens adorned in blue silk waited on him in his house in Chiang Mai.[119] (According to a more polite description, "Leonowens followed the marital customs of the land of his adoption until late in life, when he married an English girl."[120]) In 1893 Leonowens bought the Oriental Hotel from Hans Nils Andersen, founder of the Danish East Asiatic Company, after which the hotel was run as a brothel.[121] Four years later, Leonowens entered into a six-year agreement to work for the Bombay Burmah Trading Corporation, but later started his own firm. The Louis Leonowens Company had timber stations in Lampang and Muang Thon.

Registered in India with its headquarters in Bombay, the activities of the Bombay Burmah Trading Corporation—"trade with and at Burmah and its neighbouring countries"—were determined largely from London.[122] The firm entered the Siamese teak industry later than Borneo, but by 1900 became the country's largest timber firm. It made up for not having insiders like Cheek and Leonowens with large amounts of capital, which it used indiscriminately to buy up rights to harvest timber. The company obtained leases from local chiefs in the Salween forests along the Thai-Burmese border and initially resisted Herbert Slade's efforts to regulate forestry, insisting that any changes in the timber trade should be negotiated by owners, buyers, and sellers without governmental interference. But the Corporation eventually capitulated and "though it could not at the time be realized, the Corporation's future in the country was assured."[123] Bombay Burmah established stations in Chiang Mai, Lampang, and Phrae, as well as Mae Hong Son and Yuam in the Salween watershed.[124]

The Danish East Asiatic Company was founded on March 20, 1897, by the Dane Hans Nils Andersen, whose close relations with both the Thai and Danish royal households were of significant benefit to his many business ventures. With offices in Bangkok and Copenhagen, the firm was based initially on Siamese timber. Its first teak concessions were obtained at Wang Chao near Raheng in 1895. But the largest and most valuable concession was obtained in 1908 in Phrae.[125] By 1937, the company had over twenty branches and many subordinate activities, including shipping, trading, and various other industries. East Asiatic's annual average capital in the first four decades of this century was 31 million kroner with a "brilliant" profits history, averaging for almost half a century a dividend of over 14 percent per annum.[126]

The Anglo-Thai Company, originally named Siam Forest, had stations in Lampang at Ngao where the Teak Improvement Center would later be established.[127]

The sixth European logging firm, French East Asiatic Company, worked the forests around Mae Kok, a tributary of the Mekong, with headquarters at Chiang Rai.[128]

Colonial Strategies and Historical Resistance

Caú ma seì se nït' twai, hsoò hkè hpè law paw he htai.
[O, older brother, you left us to work for the foreigners for three years.
You've come home with your turban as filthy as a rag.]

Traditional Karen folk song, Chiang Mai[129]

The formation of the Thai forestry bureaucracy was a product of the political and commercial interests of Great Britain, on the one hand, and the Bangkok administration, on the other. But establishing state hegemony over forest lands meant taking rights away not only from the northern princes but also from communities living in and around the teak-bearing forests. British foresters in Siam were clearly aware that a "very considerable proportion of the many races inhabiting Siam [lived] in, on and by the forests."[130] Prior to the establishment of the Royal Forestry Department, while teak was considered the property of the northern princes, commoners were also allowed to use the smaller trees either by cutting them down or buying cut logs from others.[131] With the introduction of new regulations, villagers found their access restricted.

Since the 1980s, resistance by communities to the activities of the state and private companies has been a central beam in Thai forest politics that cannot be ignored—either by policy makers or by multinational companies planning forestry investments in Thailand. Farmers' struggles for legal recognition of customary rights to forest use and management have forced open the rigid legal edifice that ensured a state monopoly over forests for a century. How did forest-dependent people react to the centralization of control over Thai forests in the course of the past century?

In the years following the founding of the Royal Forestry Department, there were two major rebellions in the north against the general expansion of Bangkok's political power, the Phraya Phap Rebellion of 1889–90 and the Phrae Rebellion of 1901.[132] According to Chaiyan Rajchagool, the first revolt was a response to new taxation measures imposed from Bangkok, as well as a reaction to British teak extraction and the flow of timber revenues to the capitol. A key target of the second rebellion was the European and Chinese teak firms. There are other indications of foresters clashing directly not only with the northern princedoms, but with local people accustomed to having access to forest resources.

Forestry laws enacted in 1897—the Forest Preservation Act and the Teak Preservation Act—criminalized common people's use of teak by making it illegal to cut teak smaller than 2.1 meters in girth, and requiring official permits for logging teak larger than this size. As already noted, Slade's successor Tottenham alluded to an "enormous" trade in teak poles. The new Forestry Department actively suppressed people's use of teak inside concession areas by prohibiting the felling or girdling of smaller trees, while outside the logging concessions, the cutting of teak was forbidden entirely. This curtailed community forest use directly. An observer notes that in 1900, "each villager regarded any patch of young teak trees near his house as his own private property, but the efforts of the Forest Department are now beginning to convince him that such is Government property."[133] Foresters had to "explain to the astonished villagers that a teak tree was no longer the property of the first person who liked to cut it down" and that "reporting and, if necessary, prosecuting for infringements of the new rules was a natural addition to his work."

After 1913 other species would be added to the "reserved" list as they became commercially valuable.[134] Siam does not appear to have curtailed local people's customary use of forest to the same extent as in neighboring British colonies, likely because Bangkok did not have the political-economic resources to extend control to remote areas. Provisions were included in Thai forest laws as late as 1938 allowing "specified groups to collect certain forest products in restricted zones."[135] Still, 1897 marked a dramatic shift in customary forest use in Siam.

There appears to be little documentation from the early years of how "the many races" reacted to commercial logging and the gradual transfer of forest jurisdiction to Bangkok that started at the close of the nineteenth century. An anecdote told by Karen community elder Jorni Odechao, whose grandfather worked for a British timber company until 1908, suggests that rich local histories are waiting to be written.[136]

Because of their intimate knowledge of the forests and of elephants, the timber companies tended to hire Karen men and their elephants to work in the logging concessions of northern Siam. Jorni recalls that forest laborers were ordered to be careful not to let the logs crush young teak saplings when they fell and were dragged out of the forest. Otherwise, they would not be paid. Inevitably though, young trees were damaged. To hide the evidence, these were either buried or fed to the elephants. "In those days, they didn't clear-cut the way the Thai companies did. They just cut a few, only the biggest trees and left the rest," he said.

The folk song cited above, still sung in Karen communities around Mae Wang, dates from the turn of the nineteenth century. It reflects the disdain with which some villagers greeted the men who left to work for the colonial timber companies. These workers were said to earn a fair wage, but often came home empty-handed having gambled and drunk their money away. Some young women who were engaged to be married had to wait for their men for as long as seven years.[137] And when the men finally did return, they had become "old bachelors" with little to show for their absence but their dirty, sweat-soaked turbans.

Jorni also remembers being told by his grandfather that conflicts over where to log occurred regularly between British concessionaires and their Karen workers. When companies wanted to cut in forest that was considered sacred, their Karen workers bargained to have them left aside by threatening that they and, perhaps more significantly, their elephants would not cooperate in the logging. The Karen did not always get their way. Another folk song is inspired by the story of a British company that logged in the sacred forest (*de por*) where the placentas of newborns are attached to the trees, following the belief that the spirits of trees and babies are connected. One of the trees that was cut belonged to a young woman who was about to be married. When her tree was chopped down she became ill and died. In his sadness, her fiancé wrote this song:

Naù wa sei a nyà wa bau, lau kau htò hpa dö aü au.
[It is so very sad, the axe bird (*nok khwarn*) has cut her tree.]¹³⁸

In the context of contemporary Thai forest politics, Jorni's community, Nong Tao, which borders Doi Inthanon National Park, could well be prohibited access to the forests they have traditionally managed because their settlement lies outside the park's boundaries.

In neighboring European colonies, resistance to colonial forestry was fierce and varied in its expression. In British Burma, the forest-dwelling Karen presented a formidable challenge to colonial foresters' efforts to expand areas of teak extraction. The Karen adopted various strategies of resistance, ranging from flight—in some cases across the border to Siam in the decades prior to the founding of the Forestry Department[139]—to destruction of teak saplings in order to eliminate evidence of the trees in their swidden fields. Foresters realized that, as much as they would have liked, it was impossible to remove all people from all forests in Burma. As Burma's conservator, Henry Leeds, warned in 1867, "terror" would result in only limited success. Foresters' only realistic option was to convince the Karen to cooperate through the *taungya* system, where villagers were paid to plant and tend young teak plantations in exchange for temporary access to the land for growing crops between the rows of saplings. The alternative, they worried, could be much more costly: "Discretion and management will do more than all the terror which can be spread in the country. No one can judge of the mischief which people inimical to a Department which must, in a manner, be an unpopular one, can effect without being discovered."[140]

In British India, the loss of control over forests was acutely felt by communities. Foresters knew that the regions where teak and sal trees grew were not "virgin" forests. Dietrich Brandis noted that "in the wildest forest regions of India we constantly come across evidence that the land at one time had been under cultivation—fruit trees, ruins of large buildings and terraces of old fields."[141] Peasants responded to the gradual expansion of state control over forest land in various ways. Some simply fled, others held marches and strikes, some practiced "deliberate and organized incendiarism," that is, they set fire to young tree plantations, and there were occasional attacks on officials.[142]

In their field reports, foresters described the obstinacy of local people and the difficulty of imposing state control over forests. This frustration is reflected in the report of one forester in British Garhwal who wrote in 1916: "The notion obstinately persists... that Government is taking away their forests from them and is robbing them of their own property. The notion seems to have grown up from the complete lack of restriction or control over the use by the people of waste land and forest during the first eighty years after the British occupation. [They are] most assured of the antiquity of the people's right to uncontrolled use of the forest; and to a rural community there appears no difference between uncontrolled use and proprietary right... [My] best efforts however have, I fear, failed to get the people generally to grasp the change in conditions or to believe in the historical fact of government ownership."[143]

Though Indonesia was colonized by the Dutch, there, too, German and German-trained Dutch foresters established the forest service and implemented scientific forestry.[144] As in Burma and India, this meant securing state ownership of forest lands, managing timber extraction and replanting, and curtailing the customary rights of villagers to teak and other forest products. With the state's consolidation of control over the forest, the poorest villagers lost important subsistence options and found themselves squeezed off a resource on which they had been dependent. They protested by refusing to pay taxes, cut teak in open defiance of forest laws, remained on land after leases expired, and piled stones in the roads they had been ordered to build.[145] By 1928 most replanting in Java was done by *taungya* (in Java, called *tumpang sari*), and it tended to be the most disenfranchised, landless people who followed the teak concessions, planting crops for one or two years between the saplings and then moving on to the next area. As villagers resisted the state's appropriation of access rights to the forest, foresters intensified their policing role and reforestation in Java "became the art of persuading people to plant trees on state land."[146]

With a century of logging in northern Thailand, there remained very few communities that by the mid-1980s had not had timber companies on "their" land at least once. Apart from the impacts on the forest itself, people's resentment was likely fueled by the fact that profits from the teak industry were not reinvested in the areas where the resource was extracted. "Instead of building and improving roads for communities, the logging firms more often damage existing bridges and roads passing through villages," according to Banasopit Mekvichai, who further writes: "The evidence shows that the villages in the teak-bearing area . . . have derived no real benefit from the teak industry."[147] She notes that as early as 1964, the director general of forestry was concerned about poverty in villages of the teak-bearing areas and proposed that the logging companies return 10 percent of their profits to these villages to build schools, health centers, and so on. The proposal was approved in 1968, but was never effectively implemented.

Many of the villagers who voiced concern about the negative impacts of logging on their forests and fields in the 1980s had accumulated years, even decades, of experience with commercial timber extraction. The villagers in the Samoeng District of Chiang Mai Province tell of four waves of loggers that came to the forests of Samoeng during the course of the century.[148] The first was the British Borneo Company, which arrived in 1896, and extracted the largest of the teak trees. Next was the newly established state enterprise, the Forestry Industry Organization, which began after the Second World War to cut more teak, this time including the smaller trees. In 1973, when the provincial company Chiang Mai Timber received a logging concession to cut *mai krayaloei* (hardwoods other than teak[149]), the conflict between loggers and local people intensified. The company set up a checkpoint to prevent people from going in and out of the area, and for the first time a villager who entered the area to cut a single tree would be arrested.

As Nuan Lachai, a community leader in Samoeng, tells it: "The whole village was shocked. A meeting was called to discuss the matter. No one understood why the company had the right to cut down a whole forest of huge trees without getting arrested, while if one villager cut a small tree down he was. The forest had been our forest from the time of our ancestors, hadn't it? They came in and logged without asking permission from us or even telling us. We never blamed them for anything, but now we were the ones getting arrested."[150]

Attempts to organize and issue official complaints were ineffective. Part of the problem was that local officials and some members of the community were reportedly being paid off by a sawmill owner who offered protection against arrest. During the years that the Forestry Industry Organization remained in the area, villagers say the forest became "thin and mangy." They felt the spirits of their ancestors had been disturbed by logging in sacred groves (*pa cha*), and the *mueang fai* traditional irrigation system fell into disarray when branches and forest debris clogged up the feeder canals and streams.

When the fourth logging company came to Samoeng in 1986, Nuan Lachai and other members of the community realized that if their resistance did not succeed, there might be no forest left at all. They feared not only the loss of numerous non-timber products like mushrooms, bamboo shoots, and medicinal plants, but also the regular flow of water that sustained rice cultivation. This time, the villagers solicited the help of Chiang Mai University students and academics as well as Bangkok environmental groups, who put them in touch with other communities that were also demanding the cancellation of logging concessions in their own areas. They learned that in nearby Nan Province monks were ordaining trees with holy saffron robes to protect them against the loggers. They realized that as far away as Rayong in the east and Surat Thani in the south similar conflicts were brewing.[151]

Their protests contributed in no small measure to the pressure that forced the government of Prime Minister Chatichai Choonhavan to declare the world's first nationwide ban on commercial logging in 1989.

The Four Failures of Thai Forestry

Forestry in Thailand has failed. It was only sustainable in theory. After cutting, no one paid attention to replanting. We only focused on production... Even if we had followed the rules there is no guarantee that it would have worked because of the human factor.

Winai Subrungruang, Deputy Managing Director of the Forestry Industry Organization[152]

What is the track record of Thai forestry? Thai foresters have always pointed accusing fingers at others—most often the traditional enemies of their profession, the forest-dependent communities. But they have shown remarkably little inclination to examine their own role in the process of steady forest decline, insisting instead on their superior, indeed exclusive, knowledge of forests and their historic right to monopolize jurisdiction over forest lands. Has Thai forestry worked? Given the state of Thai forests today, the answer has to be "no." Until foresters reflect critically on their own history, a vital part of the Thai forest story will be missing.

Public concern about the state of the forests, expressed in the 1989 logging ban, has caused many to draw the conclusion that Thai forestry's greatest shortcoming was its failure to preserve biodiversity. This is a grave misconception. Foresters may well share the concern for biodiversity, and protecting circumscribed "virgin" areas for conservation purposes is, in principle, not disputed. But conservation was added to the foresters' mandate relatively recently. Seeing biological diversity as a value does not simply broaden the goals of forestry. It reverses them, forcing foresters to question the original, central objective of forestry science.

From the beginning, the purpose of the discipline was gradual replacement of "chaotic" and "abnormal" nature with its abundance of "weeds" and "inferior" plants with a "normal" forest, one that is increasingly dominated by one commercially valuable tree species—be it teak, pine, or eucalyptus. Biodiversity depletion was the implicit goal of the style of forestry introduced to Asia during the era of colonialism and it continues to be a necessary precondition for the establishment of large-scale monoculture plantations of eucalyptus in Thailand today. Biodiversity loss is not an unfortunate by-product of forestry; it is forestry's logical outcome. It is, at its root, a hallmark of *successful* forestry.

But acknowledging the value of biological diversity threatens forestry in another important way. It brings into focus the discrepancy between the knowledge of foresters, which concentrates overwhelmingly on very few species, and the forest knowledge of local communities, which encompasses a vastly greater part of the forest ecosystem. Compare the diversity of a well-managed tree plantation, where undergrowth is suppressed and wildlife—with nothing to eat and nowhere to hide—is virtually absent, with the 167 species of trees and plants (all with various uses), 18 species of edible plants, 15 species of medicinal plants, and 186 species of birds surveyed in one community-managed forest.[153] Forest farmers in northern Thailand not only claim that their techniques maintain biodiversity, studies show that swidden cultivation increases biodiversity, by opening up the land for grass, forest vegetables, earth mushrooms, and other types of vegetation that help to store moisture.[154] It becomes increasingly absurd to demonize people, as foresters have historically done, whose very knowledge systems reflect and maintain forest diversity.[155]

The failures of Thai forestry, then, refer not to the inability to preserve biological diversity. Rather, Thai forestry failed *on its own terms* in four important ways. It failed:

1. to limit commercial timber harvesting to a "sustainable" level,
2. to stem corruption and illegal logging,
3. to establish tree plantations that could produce timber to replace logging from natural forests, and
4. to maintain its monopoly over forest land—that is, to keep people out of the forests.

Over-cutting

Underlying the sustained-yield theory there seems to be a belief in a "golden age" (perhaps before World War II) when Thai forestry was done right. In the contemporary debate, many foresters continue to defend the theory and seem to hark back to this mythical time. Yet even official forestry statistics, which must of course be viewed with some caution, suggest that timber extraction in Thailand was *never* kept within the "sustainable" limits defined by foresters.

At the end of the nineteenth century, Herbert Slade estimated that the average teak output of the Chao Phraya basin during his time as Chief Forester was 75,000 logs per year, more than double what it should be. In his opinion, 30,000 logs was the appropriate annual output.[156] (More timber still was floated down the Salween and Mekong rivers.) When his successor, W. F. L. Tottenham, took over leadership of Siam's forestry department in 1901, the companies continued to exceed the volume of timber that the forests could sustain and unless the fellings were reduced, the existence of the forests would be jeopardized.[157] His estimate for the appropriate log output of the Chao Phraya basin had climbed just a couple of years later to 100,000; and 60,000 for the Salween basin.[158] "The forests were being worked beyond their possibilities," Tottenham wrote in 1905.[159]

In her study of the history of teak logging in Thailand, Banasopit Mekvichai notes that there was a direct correlation between the length of time a forest was exposed to commercial logging and the size of the timber. The average size of logs coming out of the Salween basin in the period 1896–1926, where harvesting by British companies had been underway since the mid-nineteenth century, was about 190 cm. In the Chao Phraya basin, where intensive logging did not begin until several decades later, the average size of timber was 230 cm. The average size of timber originating from the Mekong basin, lying even farther from Burma to the east, was 250 cm (see table 1). From 1913 to 1928, the average annual number of teak logs exploited in the Chao Phraya basin climbed to 111,631. The British advisor to the department, D. Bourke-Borrowes,[160] estimated that a "reasonable output" should have been just 50,000 logs per year (60 percent higher than Slade's calculation thirty years earlier).[161] He predicted a "depressing" future, warning that by 1960 Thai forests would only be able to produce half this amount. The truth, in a sense, turned out to be worse. From 1900 to 1960, Thai forests were stripped of the largest, most valuable teak trees. In terms of the size of timber, the average volume of logged teak dropped by more than half from around 2 cu m per log at the beginning of the century to just 0.63 cu m per log in 1960.[162] Over the same period, the average girth (circumference of the bole at breast height) of teak declined dramatically from over 2 m to a predominance of the smallest exploitable size of 90 cm to 150 cm.[163] In Slade's time the *minimum* exploitable girth for teak had been 213 cm. By 1960, when the last foreign concessions were cancelled, the standard had been adjusted downward to just 100 cm.[164]

Banasopit notes that the foreign companies "took an enormous amount of timber out of Thai forests shortly before the expiration of their concession leases (many of them in 1955), as they attempted to earn as much from the concessions as they could before they had to leave."[165] Indeed, her data shows that total teak production jumped from about 240,000 cu m per annum between 1948 and 1952 to well over 300,000 cu m in 1953, 1954, and 1955. Over the next twenty years, annual teak production volumes (legal plus confiscated illegal timber) hovered at around 180,000 cu m.[166] When the Food and Agriculture Organization (FAO) conducted its inventory of Thai timber resources in 1958, it concluded that the marketable teak stock of five northern provinces—Chiang Mai, Lamphun, Lampang, Chiang Rai, and Phrae—was 2 million cu m. This, according to FAO, was just 20 percent of what had been available two decades before.[167]

As for *mai krayaloei* (hard woods other than teak) the production volume increased steadily from about 1.1 million cu m in 1950 to 1.5 million cu m in 1958. As the last European timber companies' leases expired, the volume logged dropped suddenly back to 1 million cu m in 1959, after which it proceeded to rise gradually, reaching an unprecedented 3.2 million cu m in 1977.[168] In January 1979 the Thai government prohibited logging in half the concession areas in an attempt to preserve rapidly dwindling supplies. By 1985 only 13 percent of the 196,000 sq km designated as timber concession areas were being worked.[169] When the logging ban was declared four years later, the total reported teak output was down to 47,000 cu m and only eleven teak concessions were still under operation, the rest having been exhausted of harvestable timber.[170]

Table 1 Thai teak logging, 1896–1981

Year	Avg. volume/log (cu m)	Avg. girth (cm)	Legal output (cu m)	Reported illegal output (cu m)	Illegal output (% of total)
1896–1926	2.11 (Chao Phraya) 1.34 (Salween) 2.58 (Mekong)	230 190 250	224,356	–	–
1932–36	2.13†	230†	195,171	–	–
1937–41	1.30‡	190‡	140,128	–	–
1947–51	1.05§	160§	206,392	–	–
1952	0.96	150	243,185	18,121	6.9
1953	0.82	140	310,557	35,399	10.2
1954	0.78	140	306,927	51,951	14.5
1955	0.77	140	256,456	49,419	16.1
1956	0.75	130	151,981	48,314	24.1
1957	0.80	140	146,670	41,021	21.8
1958	0.79	140	127,306	53,955	29.8
1959	0.68	130	55,726	41,202	25.2
1960	0.63	120	77,823	75,841	49.4
1961	0.63	120	57,310	48,355	45.8
1962	0.63	120	89,046	34,278	27.8
1963	0.79	140	121,686	22,291	15.5
1964	0.62	120	104,518	38,681	27.0
1965	0.60	120	146,116	107,558	41.9
1977	0.64	120	112,210	25,761	18.7
1978	0.26	90	73,486	38,724	34.5
1979	0.59	120	146,646	32,931	18.3
1980	0.51	110	74,444	22,879	23.5
1981	–	–	65,767	7,478	10.2

† For 1935, ‡ For 1940, § For 1951

Source: Adapted from Banasopit Mekvichai, *The Teak Industry in North Thailand*[171]

Corruption

On an October morning in 1999, then director general of the Royal Forestry Department Plodprasob Suraswadi grants an interview in his spacious office atop the gleaming new H. A. Slade Building. Asked why sustained-yield forestry never worked in Thailand, he answers emphatically and without hesitation: "Corruption! They only cut the good trees without replanting... they didn't follow the rules... people cheated!"[172]

In the version of the Brandis system that was practiced in Thailand during most of the last century, a logging concession was divided into thirty compartments, and timber was extracted from each one over a thirty-year period. By the thirty-first year, a concessionaire was supposed to be able to return to the first compartment where more trees were to have reached a harvestable size. Thailand's teak expert Apichart Kaosa-ard explains that in theory "thirty years was the time required for 'second class' trees to replace the older ones that had been cut."[173]

Though 100 cm was the minimum girth for teak, not all trees of this size and greater were to be cut. It was, for example, forbidden to cut the tallest, largest, straightest trees because these should be left as "mother trees" to reseed the area. Trees along riverbanks were to be spared because they helped to maintain water courses, and ridge trees on steep slopes were to be left as they protected against erosion. Of the remaining trees of the appropriate size, only 20 percent were to be felled during one rotation. Moreover, the most valuable timber from among this 20 percent was not to be cut. Rather—and this was key in the concept of improvement—a *random* selection was supposed to be made.[174] If anything, the "bad" and diseased trees were to be taken out first so they would not "contaminate" the forest with their inferior seeds.[175] These rules were designed to enhance the seed stock of teak in each compartment through successive generations of harvesting. Clearly, they required that loggers exercised self-control of heroic proportions.

But even under ideal conditions, the system had its problems. According to British forestry historian H. Colyear Dawkins, the "greatest complication" with the Brandis system arose from the fact that "some of the trees over the minimum felling diameter are very large and very old—much older than the age of the largest trees that will be cut in the second cycle. If all these trees are removed in the first cycle, then the yield in the second cycle will, inevitably, be less." This problem arose in "country after country, and forest after forest, time and time again, [throughout the twentieth century]."[176] Apichart describes a similar process in Thailand: "The Brandis system became 'negative selection' because the best trees tended to get cut."[177] The tall, straight trees and the ones growing along riverbanks were the most attractive and easily accessible. The steepness of a hill was no obstacle for elephants that can go almost anywhere in the forest. If there were fifty harvestable trees in a compartment, only ten were supposed to be cut, but this limit was often exceeded. If, as might be the case given the natural uneven distribution of teak, the first compartment did not have enough harvestable timber at all, it was not uncommon to dip into the second.

Selection, moreover, was a precarious business. Badly underpaid foresters reportedly received a flat-rate payment from companies for each tree selected. The income of the "selector" was thus directly dependent on the merchants, who would only pay for commercially valuable timber. Both foresters and merchants lost income if "bad" trees were cut. In any case, in reality, foresters were often not the ones who did the selecting. An early description of how things actually worked is provided in Boonsong Lekagul's treatise on Thai forestry from 1959, *Hak pa mai yang yu yang yuen yong* [If the Forest is to Last Forever].[178] "As any Thai forester will tell you, it was often the concessionaires who went out to the forest, chose the trees they wanted, logged them, and dragged them out of the forest. Only then did foresters place their marks on the logs."

Cheating within the concession system was only part of the problem. First reported by Thai foresters in 1937, illegal logging (outside the concessions) had by the end of Thailand's logging era "become more widespread than legal logging . . . [constituting] one of the major reasons for the degradation of the teak forests in Thailand."[179] This was not possible without the support, tacit or otherwise, of state foresters.

Official forestry statistics show that reported illegal logging peaked in 1960, constituting about half of total teak output. Through the 1970s, reported illegal teak logging made up for between 20 and 30 percent of the total, while unreported illegal logging was certainly much more.[180] Some measures to deal with the problem only made things worse: The Royal Forestry Department offered rewards for returned stolen logs, which created an incentive to steal timber and then turn it in and collect the reward.[181] Meanwhile, confiscated logs were auctioned off by the Forestry Industry Organization, generating 20 percent of the agency's annual income.[182] Illegal logging peaked again after the logging ban when confiscated teak logs made up 60 percent of total output and confiscated timber of other reserved species made up 150 percent of the total.[183] Through the 1990s, illegal logging, as one would expect, exceeded legal logging. However, as timber production in neighboring countries soared, total Thai timber production dropped from over 2 million cu m in 1987 to 119,000 cu m in 1992 to 54,000 cu m in 1998.[184]

Banasopit Mekvichai divides illegal logging during the concession era into three types. The least significant of these was performed by villagers acting on their own for their own consumption. The second type involved villagers working in groups with the financial assistance of a "forest capitalist." Here, it was the villagers who took the risks and were paid very little for their work, in comparison with the value of the logs. If they were arrested, their "bosses" bailed them out of jail.

In contrast, the third and most significant form of illegal logging involved teak merchants hiring groups of outsiders to log, frequently antagonizing and threatening local people to ensure their silence about the illegal operations. "The logging operations done by groups of outsiders working directly for teak merchants are on a much larger scale and have often led to violence when confronted by forestry officials and police not bribed by the teak merchants," Banasopit wrote in 1988.[185] Equipped with trucks, chainsaws, tractors, and elephants, they resembled the legal operations "except that they must be smaller in scale to elude the government officials who are not in collusion with them." To succeed, the teak merchants had to "bribe many of the lower-ranking forest officials who are responsible for protecting the forests."

Of course, not all forest officials were corrupt. But they were underpaid and faced an almost impossible battle with illegal loggers who could use all manner of tactics to hamper the foresters' work, from placing nails in the road or rolling trees to block the way, to outright guerilla tactics with automatic weapons. If arrests did take place, bail was quickly paid by the teak merchants.

Even after the logging era, being an honest forester continued to be a dangerous, hopeless business. In April 1993 the newly appointed forestry chief for Ranong Province, Pheerasak Adisornprasert, sent out a clear message to timber poachers by halting the illegal activities of one local businessman who had encroached on public forest land. A few days later, Pheerasak was gunned down in broad daylight by an assassin. According to then deputy agriculture minister Suthep Thaungsaban, the killer had been hired by a "local influential figure . . . a former government official with strong ties to politicians."[186]

Plantations

Logging was unsustainable and corruption was rife. Had Thai foresters managed to build up tree plantations though, they might at least have been able to ensure a supply of timber to replace logging in natural forests. But plantations were never given serious priority. Though replanting was required of both, neither foresters nor concessionaires made the necessary effort or investment. After eighty years of planting, Thai forestry authorities had established tree plantations in an area equivalent to the average amount of natural forest that disappeared each year during the 1970s and 1980s.[187] According to Forestry Department statistics, a century of state reforestation efforts resulted in plantations covering only 1.7 percent of the country's surface area.[188]

The first trial plantations of teak were established by the Royal Forestry Department in 1906 in Phrae Province. By then there was already a rule in place requiring concessionaires to plant four trees for every one they cut.[189] The companies' failure to comply led in 1924 to the introduction of a "forest maintenance fee" that came on top of the royalty payment for each concession area. This was to be used by foresters to plant teak in plantations.[190] But its effect was only to take the pressure off companies and to increase the department's income without resulting in a corresponding expansion in plantation area.

The British forestry adviser Bourke-Borrowes pointed out in 1928 that "the small trained staff was usually so completely immersed in revenue collection that forest survey work of any kind is of the rarest occurrence... No regular program of silvicultural work has ever been drawn up and carried out in Siamese forests."[191] Moreover, he wrote, there were no work plans, no maps, no annual reports, little improvement felling, an "almost complete absence of scientific data," and only 1.2 sq km of teak plantations had been established. Bourke-Borrowes concluded that all in all, operations were of a "meagre, intermittent and spasmodic nature."[192] Up to the 1960s, the area of planted teak rose to 80 sq km.[193] By 1986 it had expanded to 1,330 sq km, much of it planted by the Forestry Industry Organization.[194] Eight years later, the total area of tree plantations other than rubber was reported to be over 8,000 sq km, 1.7 percent of the country's surface area.[195]

While Bourke-Borrowes acknowledged that the ecology of Siam was "very complicated and difficult,"[196] he described Siamese forestry as hopelessly backward in comparison with the forestry practiced in India and Burma, where the state had real control over land and forest.[197] This kind of argumentation mirrored the claims of British foresters in India and Burma, who complained of the inferiority of forestry practices in the Asian colonies compared with that of Germany and France.

Indian and Burmese forestry were the prototypes on which Herbert Slade based his proposals for Siam. Yet in 1900, Slade's fourth year in the kingdom, Inspector General of Indian Forests R. Ribbentrop wrote that "the science... of the correct treatment of all classes of Indian forests [was] as yet in its infancy." Instead of creating even-aged plantations as had been done in Europe, it was "necessary to rely almost entirely on the natural reproduction of our forests. For a more intensive management, the areas to be treated are by far too vast."[198] Sixteen years later, the situation had not improved much. Ribbentrop's assistant, Robert Scott Troup, wrote that Indian and Burmese forestry consisted merely of felling "every mature teak tree which [can] reasonably be cut out from amongst the mass of miscellaneous species in which the teak is scattered."[199] This "*quasi*-selection system" failed to transform the chaos of Indian or Burmese nature into a "normal forest" and differed "in the most fundamental principles from the true selection system of Europe."[200]

In 1921 Troup, who had by then replaced Wilhelm Schlich as chair of forestry at Oxford University, stated that the scientific treatment of the "large areas of irregular and for the most part mixed forests of India and Burma was one of the most difficult problems with which forest management has been faced ... the constant removal of more valuable species must inevitably lead to the deterioration of the forest."[201]

Thus in this absurd hall of mirrors, Thais were told they had to mimic forestry in India and Burma. Indians and Burmese were taught to emulate the forestry systems of Central Europe. German foresters, meanwhile, would only confront the ecological and economic costs of a style of forestry that became the global model at the very end of the twentieth century.

But Thai plantations failed not only because of the way trees were managed. Another reason that the planting proceeded so slowly was that it relied on exploited laborers who had no interest in the success of the endeavor. While ownership of plantation areas rested entirely in the hands of the state, the onus of the work was placed on local people who received very little in return. In many cases, they also lost agricultural or forest land to the teak plantations. They were hired to do the work according to a version of the Burmese *taungya* system, in Thailand referred to as "forest villages" (*muban pa mai*). According to Amnuai Corvanich, former managing director of the Forestry Industry Organization, labor for the plantations was obtained by two members from each family being given the responsibility to plant teak on 10 rai each year (that is, 5 rai planted by each worker).[202] They were paid 200 baht per rai for a total of 2,000 baht per family per year. They were also permitted to plant their own crops between the teak trees for the first few years and to grow crops on another 5 rai of land allocated to each family.

A direct legacy of colonialism, the *taungya* system assumed that villagers would plant and tend teak in exchange for temporary access to land. Each FIO forest village was supposed to produce 10,000 rai (16 sq km) of new teak plantations every ten years. The agency had jurisdiction over a total potential teak planting area of 5,760 sq km, but managed over a period of thirty years to establish only 924 sq km of teak plantations—of highly varying quality.[203]

The forest village system failed because of the abusive terms of the *taungya* system. The salaries paid to laborers were such that "families who remain on the plantations often supplement their incomes through illegal logging . . . Instead of protecting the teak trees . . . they are forced by their poverty to fell the trees so they can earn enough for their families to live."[204] A former director general of forestry, Dusit Banijbatana, commented that "most of the benefit [in the *taungya* system] goes to the owners of the forest. The benefit to the farmers is only in allowing them access to agricultural land which they are not considered to own."[205]

Former deputy managing director of FIO Winai Subrungruang was, as a young forester, responsible for implementing FIO's forest village program starting in 1969. Looking back, Winai offers a surprisingly critical assessment of the program: "The weakness of our *taungya* system was that villagers got nothing after the planting period was over. We just saw them as laborers. We based this on the Burmese system. But we realize now that members of these forest villages were being exploited. They got nothing out of it!"

The result was not only that teak plantations were badly taken care of. They often became the target of people's wrath. Winai reports that "people encroached on our plantations, destroyed the saplings, burned them, and pulled them out."[206]

Policing and encroachment

Forestry's obsession with removing people from forest land has often been depicted as a response to the increasing pressure placed on a limited resource by growing rural populations—whether tribal hill people or ethnic Thai lowlanders. This is just plain wrong. Limiting local people's access to forests as a key strategy of foresters pre-dates by a century Thailand's rapid population growth that followed the Second World War. Eviction was never a secondary issue for forest managers who otherwise spent their time tending trees. On the contrary, asserting full state control over forest land was central to the science of forestry from its colonial beginnings. Forestry has, in this sense, always been about evicting local people.

Colonial governments in India and Burma did not declare exclusive rights to forests *in spite of* the presence of communities; they did so, at least in part, *because* people lived in and claimed rights to these lands. Embedded in the notion "forests are state land" is an implicit rejection of customary rights, an assumption that the people living there and using the resource must be removed. Yet foresters' classical training focused exclusively on a very few species of trees and taught them nothing about how to deal with rural communities. Removal was literally an impossible task for which foresters in Thailand, and perhaps all tropical countries, lacked the budget, manpower, time, and skill. Still, the policing role of foresters since the nineteenth century has been a key element of their mandate.

As we have seen, there did exist debate among colonial foresters about the impact of swidden agriculture on tropical forests with some suggesting that this method of burning and leaving fields fallow might actually have a positive effect on teak timber.[207] Indeed, Thailand's first Chief Forester Herbert Slade fundamentally questioned European foresters' uncompromising view on annual fires set by forest-dwelling people. The majority view, however, was that fire destroyed trees and all burning was therefore bad. Though adamant about the important role of fire, Slade maintained that shifting cultivators were causing untold damage to Siam's forests. Arguing along similar lines as Dietrich Brandis two generations earlier, Slade noted that in Siam "one could wish to entirely stop this ancient but wasteful practice" and to compel people to confine themselves to certain areas. But the issue "must be attacked with tact and discretion or there will be trouble with these worthy but half tamed people. The damage they do now is incalculable and they annually destroy many square miles of teak forest."[208] Setting the tone for a century of Thai forestry, Slade declared that the most important task for his successors would be the "regulation of hill clearings."[209]

For a century then, Thai foresters have had the responsibility to remove people from forests, but have never had the necessary tools. The years passed and the number of forest "encroachers" grew. By 1974 an estimated 5 to 6 million people were living illegally as "squatters" on untitled land in forest reserves.[210] By the late 1980s, it was common in the forest debate to speak of 10 million people living in forest lands; that is, a quarter of the rural population.[211] According to the Regional Community Forestry Training Center for Asia and the Pacific (RECOFTC), the number had reached 14 million living "in or near" forest land by 2006.

What to do with so many people became an overwhelming concern of Thai forestry. Removal as a strategy had failed utterly. The 1985 National Forestry Policy was in this sense a declaration of surrender. It proposed to get rid of the problem once and for all, not by removing people, but by transferring the costs and political headaches of dealing with landless farmers from the state forestry administration to private forestry companies. It stopped short of de-gazetting state forest lands, and instead offered firms long-term leases with negligible rental fees. But peasant resistance stalled several attempts to establish tree plantations in forest reserves on the massive scale envisioned by the policy. Widespread protests even led to the cancellation of the military's resettlement scheme, *Khor Chor Kor*, which sought to use sheer force to remove landless people from state forest land to make way for industrial tree farms.

By 1992, with commercial logging banned and large plantations stalled, it had become impossible to justify the Forestry Department's jurisdiction over such an immense, inhabited, and only partially forested area. The Ministry of Agriculture undertook a survey whose aim was to determine how much land was really forest and how much was farmland. It concluded that of the total 235,200 sq km of legally designated forest (45 percent of the country), 11,200 sq km had already been converted to farmland and 59,200 sq km were too degraded to be considered for conservation purposes, a total of 70,400 sq km or about 13 percent of the total land area.[212] This provided the basis for a crucial cabinet decision taken on September 14, 1995 to transfer jurisdiction over these 70,400 sq km from the Royal Forestry Department to the Agriculture Land Reform Office (ALRO).[213]

According to forester Krishna Brikshavana, most of the country's landless people lived in these areas. This simple administrative move cut the area in half for which RFD was responsible, and set in motion a major redefinition of the department's mandate. It effectively washed foresters' hands of most of the intractable problem of landless people, and turned their primary focus to "good" forest. Equally important, a new possibility opened up for private forestry companies to bypass the forest bureaucracy and obtain degraded state forest land directly from ALRO, according to a long-term leasing arrangement termed "contract farming."

Though the primary goal of foresters was now radically changed from managing logging and tree farms to protecting natural forest, their historic preoccupation with removing forest occupants and securing monopoly jurisdiction over forest land remained.

The Seventh National and Economic Social Development Plan (1992–96) reversed forest cover targets specified in the National Forestry Policy of 1985 to 25 percent conservation area and 15 percent economic forest. With no more (legal) commercial logging, and relatively little involvement in plantations, the task left to foresters was to guard and expand the boundaries of existing protected areas. Finding out exactly how many people were living inside them suddenly gained new urgency. How many people inhabited this 25 percent (140,800 sq km) of "good forest"?

Curiously, given forestry's obsession with illegal encroachers, no census of people living inside forest land had ever before been carried out by the Royal Forestry Department. In 1998 forester Krishna Brikshavana was ordered to undertake such a survey in areas classified as conservation forest. By 1999 he had tallied 244,212 households.[214] Subsequent surveys increased this number to 460,000 households, suggesting that the number of people living inside conservation areas could be well over 2 million.²¹⁵ Many of these people are ethnic hill people, and the inevitable clashes and conflicts in protected areas between foresters and local people have in recent years taken on an ugly racist tone.

PART 3

THE LOGICAL CONCLUSION: FACTORY FORESTS

Forestry professors have only been thinking about the commercial production of teak, pine, and eucalyptus. This is why forestry science in Thailand is so limited, so backward, so uncreative.

Veerawat Dheeraprasart

In the era of concession forestry, the main activity of Thai foresters was to monitor timber companies and collect revenue. Replacement planting, which had been mandatory since 1906, was never taken seriously either by the companies or by the forestry authorities. The government department that was established to ensure the long-term extraction of timber from natural forests failed to fulfill its mandate, and when logging concessions were canceled nationwide, Thai foresters lost their *raison d'être*. Since the ban, they have faced an existential dilemma, and have been searching for a new mandate for the post-logging era. Short of a thorough rethinking of the purpose of Thai forestry, two options emerged for foresters: (1) facilitating commercial tree planting in degraded land, or (2) assuming the role of biodiversity guardians in the country's remaining forest. This section deals with the first option; part 4 examines the second.

Classical forestry in India, Burma, and Siam concentrated mainly on extracting the best of the teak that occurred "naturally" and attempting to "give natural reproduction the best possible chance" through planting, weeding, fire prevention, and limiting local people's access.[1] The ideal of foresters, though one it never really attained, was a "normal" forest where teak trees of the same age grew—if not in rows, then at least clumped together in extended loggable areas.[2]

Industrial forestry takes this goal of normalcy much farther. Here, it is not simply about planting trees of the same species and age in even rows on a large scale. It is about transforming the genetic makeup of trees both to enhance their timber-producing qualities and to ensure their uniformity. Modern forestry pushes the old ideas far beyond what nineteenth-century foresters could have dreamed was possible—toward their logical conclusion, a stand of clones, a "factory forest" with a genetic diversity of one.

Conceptually, foresters have operated within the bounds of two related ideas. One is that the "stability of a community and its constituent species is positively related to its diversity—the more diverse, the more stable and hence the less likelihood of destruction."[3] The corollary of this general ecological principle is that "everything planted in monoculture is more vulnerable to insects, disease, fire, and storm than diverse systems."[4] Yet the oxymoronic notion of a monoculture forest has been precisely the holy grail of the forestry practiced in Thailand during the last century, most intensively pursued since the 1980s with the introduction of eucalyptus. The paradox of this type of forestry is that it relies on the broad array of genetic choices contained in natural forests to enable a radical narrowing of the genetic makeup of trees. Modern industrial foresters, like other genetic modifiers, need the natural, diverse, "abnormal" forest to supply raw material for industrial tree plantations. Artificial homogeneity depends on the survival of natural diversity. Moreover, unlike agricultural crops that can be harvested within a year and whose genes can, it is argued, be preserved *ex situ*, the long growing time of trees means that genetic diversity can only be maintained in natural forests.

In Thai forests that diversity includes some three thousand tree species.[5] Yet Thai forestry has for a century focused overwhelmingly on just three: teak, pine, and eucalyptus, only two of which occur naturally in Thai forests.[6] When we speak of the "timber bias" in Thai forestry, it is these three types of trees that we mean.

How did a century of forestry in an immensely diverse ecosystem end up with such an excruciatingly narrow focus? There are of course many reasons for this. These three species have dominated tree plantations throughout the tropics, and their prevalence in Thailand is, in some senses, merely a reflection of a global forestry phenomenon. But in the

Thai context, it is impossible to understand how Thai forestry came to virtually ignore 99.9 percent of the forest's trees (not to mention the rest of the ecosystem) without considering the curious role of the Danes. The point here is not to blame little Denmark for Thailand's forest crisis. Yet this far-away Nordic kingdom played a crucial part in the Thai forest story, and any telling of that story is incomplete without a Danish chapter.

The Danish Factor

Though a small country with few forest resources of its own, Denmark played a crucial role in narrowing the focus of Thai forestry. While many Western donors have throughout the development era been involved in Thailand's forestry sector, it is the Danish intervention that has the longest and most varied history in Thai forestry.[7]

During the first half of the twentieth century, Danes logged teak. For most of the second, they supported research on teak and pine seed "improvement." This, in turn, led directly to the identification and promotion of the non-native *Eucalyptus camaldulensis*, which eventually became the pulp and paper industry's species of choice and the bane of landless farmers. A combination of Danish commercial interests in the teak business, colonial politics, and the personal influence of two Danes—a former owner of the Oriental Hotel and a world-famous forest geneticist—laid the foundations for the thinking that lies at the heart of plantation forestry in Thailand. The fact that tree planting in Thailand largely failed does not reduce the significance of the Danish involvement.

Danes often say half-jokingly that Thailand was the "most successful colony they ever had" because of the extensive Danish involvement in Thailand over the last century and a half. The phrase is a half-joke because Denmark was, of course, never a major colonial power in Asia. A history of Danish-Thai relations published by the Danish East Asiatic Company explains that "unlike other western powers, Denmark had no interest in acquiring colonies or expanding her influence through religion."[8] This was, however, not for lack of trying. Starting in the late 1600s, the Danish Royal Asiatic Company attempted on several occasions to establish trading posts in the Spice Islands (Indonesia), Ceylon (Sri Lanka), and India. Over and over, the Danes were outmaneuvered by the wily Dutch. In contrast, Denmark was much more successful trading African slaves, rum, and sugar between the Gold Coast (Ghana), the Americas, and Europe.[9] But by the late 1800s, Denmark had neither the economic nor the military power necessary to establish large Asian colonies.[10]

Nevertheless, in Siam, Denmark found a niche. King Chulalongkorn, who faced the very real prospect of annexation by Britain on one side and by France on the other, appointed Danes to key positions as a way of limiting the country's dependence on British advisors. As such, Danes became involved in developing a vast number of state activities in Siam, including the navy, provincial police, army, tramways, telephone lines, postal service, electricity system, cement production, and railways.

Danes in Siam

Captain Andreas du Plessis de Richelieu was among the fifty Danes who served in the Siamese navy. He worked there for twenty-six years and eventually became naval commander-in-chief in charge of the Siamese fleet.[11] The precursor of the Thai Border Police, the Department of Provincial Gendarmerie, was established in 1897 with the Danish Captain Gustav Schau as its first director. Other Danish officials who worked in the department included Colonel Palle Warming, Major Erik Seidenfaden, and Captain Michelsen,[12] who died fighting for Siam at Phrae in a battle against the Shan.[13] In the 1880s, a Danish company was given a concession to construct the first railway line between Bangkok and Samut Prakan, the "Pak Nam route." The Dane Hans Nils Andersen, who went on to found the Danish East Asiatic Company, was in charge of the Bangkok-Ayutthaya line, as well as lines to Nakhon Ratchasima and to Kanchanaburi. The Siam Electric Company was a Danish-owned company that had a contract to develop

> an electricity system for Siam.[14] The Siam Cement Company, registered in 1913, ordered its equipment from Denmark and was managed by Danes until well into the 1950s.[15] Moreover, one of the founders of the Siam Society in 1904 was the same Erik Seidenfaden who was elected honorary president in 1924 and became its first lifetime member. (His nephew would play a crucial role in bringing Danish forestry expertise to Thailand thirty years later.) The new library of the Society was funded in part by Siam Cement, Danish East Asiatic, and other Danish companies.[16]

Hans Nils Andersen

Crucial to the story of Danish influence in Thai forestry thinking is the Danish East Asiatic Company, one of the six European firms logging teak in northern Siam. The firm was founded by the Danish sailor Hans Nils Andersen who settled in Siam in 1879. In his rags-to-riches story, the son of a drunken father and a mother who was forced to do the most menial jobs[17] became "influential in the development of the close ties between Denmark and Thailand,"[18] trusted equally by both royal courts. Andersen lived at first among the community of Danish sailors who worked on British sailing ships, making Bangkok their home since the 1850s. Andersen's first firm, Andersen & Co., was housed in the Oriental Hotel, which he owned from 1886 to 1893. After arriving in Bangkok, Andersen quickly developed friendly personal relations with members of the Thai royal family. In return for overseeing the construction of ships for the Siamese navy on behalf of King Chulalongkorn, he was awarded logging concessions in the north. The first teak concession was at Wang Chao near Raheng in 1895, while the largest and most valuable concession was obtained in 1908 in Phrae. The firm also had concessions for non-teak species in the Bangkok area, in connection with the construction of railway lines. During the first twenty years of its existence, timber thus provided the main income for Andersen's company, which became a major international concern.[19]

Throughout the century, the Danish East Asiatic Company enjoyed a special status in Thai-Danish relations. Indeed, the lines between commercial enterprise and state diplomacy often blurred. Members of the East Asiatic's board included prominent Danes such as Andreas Richelieu (commander of the Royal Siamese Navy) and Johan Henrik Deuntzer who would later become Danish prime minister. Denmark sided with Siam when French gunboats arrived at Paknam, and Andersen was so trusted by the Siamese Court that King Chulalongkorn sent him on a diplomatic mission to Europe in 1893 to negotiate with France.[20] In 1898 the king appointed him as the first Siamese consul to Denmark.[21] But he moved in royal circles in both countries. Just fifteen years later, he was acting as King Christian's main foreign policy advisor and "Denmark's unofficial Foreign Minister."[22] In 1921 Prince Axel joined the Danish East Asiatic Company and served as managing director from 1934 until his retirement in 1953.[23] He replaced Andersen as chair after his death in 1938, and held the position until his own death in 1964.[24] In 1930, the Danish Crown Prince Frederik traveled to Siam aboard an East Asiatic liner and delivered a personal invitation from Andersen to Prince Damrong to visit Denmark. Prince Damrong did so later the same year, along with his two daughters. All royal visits to either Thailand or Denmark, it seems, included courtesy calls on the offices of the Danish East Asiatic Company in Bangkok and Copenhagen.[25] The manager of the Danish East Asiatic Company in Bangkok has served as Danish consul general since 1927, and though these functions have become limited,[26] the consulate's emblem still hangs in the company's thirty-fifth floor office on Rama IV Road.

It should come as no surprise then, given Andersen's teak interests and King Chulalongkorn's habit of appointing Danes to key government positions, that a Dane was chosen in 1895 to reform Siam's timber business. But for a bout of cholera, Siam's first conservator of forests might not have been British, Denmark's fingerprints on Thai forestry might have been even more pronounced, and the Thai director general of forestry might today have his office not in the H. A. Slade Building but in the J. Castenskiold Building.

Jørgen Castenskiold (1861–95) came to Bangkok in 1885. He was the son of the Lord Chamberlain of the Danish Queen Louise and had on his arrival the rank of second lieutenant. Castenskiold joined the "circle of Danes" around King Chulalongkorn. He was first employed by the Siamese army as a captain in the King's Guard until 1889,[27] and then joined the Danish East Asiatic Company working in the timber trade.[28] According to James Ramsay, Castenskiold was originally hired by the Bangkok government to deal with the court case concerning the American doctor Marion Alonzo Cheek, which had become a politically sensitive and protracted legal battle between the Siamese and American governments. As that dragged on, Castenskiold became an expensive burden, and hiring him to put order in the teak business, of which he had some knowledge, removed him from the Cheek affair. In 1895 he was given the task of surveying the timber resources of northern Siam with a view to establishing a European-style forest management system. The plan then was to send him for forestry training in Dehra Dun when he returned.[29] But Castenskiold died in Tak before he could complete his mission and Herbert Slade, as we have seen, took over.

From logging to seeds

Following the Second World War, the timber concessions of the East Asiatic Company, the Bombay Burmah Trading Corporation, and other European firms were canceled by the Thai government, and handed over to state agencies and local entrepreneurs. But East Asiatic again played an important role in the next phase of Danish involvement in Thai forestry. This took the form of a botanical survey that began in 1955.[30]

The 1955 survey was initiated by three Danes: the newly appointed Danish ambassador to Thailand Gunnar Seidenfaden (an orchid enthusiast who happened also to be the nephew of Erik Seidenfaden, a cofounder of the Siam Society), and by two employees of the Danish East Asiatic, Erik Jansen and Torben Brüel, who worked the company's last teak concessions in Chiang Mai until they were canceled in 1959.[31] Prince Axel, then managing director of the East Asiatic Company, secured Danish funding for the survey through the Danish Expeditionary Fund whose main source of financing was, again, East Asiatic. This consisted of a number of excursions starting in 1957. The following year, a larger expedition was organized resulting in the collection of eight thousand specimens. This reportedly inspired other countries to get involved, and led to similar expeditions funded by the Dutch and Japanese governments.[32]

Seidenfaden was joined on the first survey by the renowned Danish forest geneticist Carl Syrach-Larsen, a founder of the population improvement theory of forest genetics, who had for years been a member of the East Asiatic's Plantation Board.[33] While Seidenfaden's passion was orchids, Syrach-Larsen's focus was teak. His arrival in Thailand coincided with the English-language publication of his seminal *Genetics in Silviculture*. It was here that the Danish forester laid out his argument for a new kind of forestry based on tree breeding to obtain individuals with desirable characteristics that could then be cloned, multiplied, and planted on a large scale.[34] In one of his first reports from Thailand, Syrach-Larsen noted that the country's forests had thousands more species of higher plants than Denmark. But teak was "by far the economically most important tree of the many species found in Thailand." It was therefore not surprising, he wrote, that the surveyors concentrated on this species alone.[35]

The botanical survey laid the groundwork for the next forty years of Danish support to forestry in Thailand, which, like the survey, was inspired by the commercial importance of teak. A history of this period attributes this support to concern over the country's disappearing forest resources and the need to maintain the timber industry: "The main reason for beginning teak breeding was the need to replace logging in natural forests with logging in plantations."[36]

Official Danish support began in 1965 with the establishment of the Teak Improvement Center at Ngao in Lampang. This was followed five years later by a Pine Improvement Center in Chiang Mai, also funded by Denmark, where trials of the native *Pinus kesiya* and *Pinus merkusii* and of the exotic *Eucalyptus camaldulensis* were first carried out. In addition, the doctoral studies of several influential Thai foresters were financed by the Danish government. The first of these was Tem Smitinand, Thailand's best-known botanist and former dean of the Kasetsart University forestry faculty, who was sent to study forest botany in Denmark in 1957. Sa-ard Boonkird, a member of the first botanical survey with Seidenfaden and Syrach-Larsen, followed three years later.[37] Others included Apichart Kaosa-ard, now a world teak expert, who took over leadership of the Teak Improvement Center following his graduation in 1979 and continued to lead the organization until 1994. Reungchai Pao-sujja, the "godfather of *camaldulensis*,"[38] studied forest genetics with Syrach-Larsen's American colleague Brian Zobel in Southern Carolina and headed the Pine Improvement Center upon completion of his studies.

While the Danish influence on Siam is clear, Thailand also played an important role in the development of Denmark's international presence. Danish International Development Assistance (DANIDA) was established in 1963, with Thai forestry support among its first projects. The Thai-Danish dairy project was another. International interest in Danish teak breeding in Thailand resulted in Denmark being invited to join FAO's Panel of Experts on Forest Genetic Resources established in 1968.[39] This in turn led to the creation of the DANIDA/FAO Forest Tree Seed Centre in Humlebæk, supported entirely by Denmark with the name: Danida Forest Seed Centre.[40] Its priority species was until recently teak. The goal of the center was to study variations among populations throughout teak's range, explore its potential for plantations in various climatic conditions, and determine the best seed sources. The center has helped to spread teak to thirteen countries in Africa, Latin America, and Asia.[41]

But in the late 1980s, public concern about biodiversity loss began to make inroads into the European forestry world, and forestry's contribution to the biodiversity crisis was increasingly recognized, even by foresters. Denmark's own forestry tradition, founded by the German forester J. G. von Langen in the late eighteenth century, was based on the establishment of plantations of faster growing, non-native species in hundred-year rotations.[42] As with Germany, extensive damage caused by windstorms tipped the scales in the forest debate causing Danish forest policy to shift in the 1990s from one based on monoculture plantations of exotics to a more "close-to-nature" forestry that mimics the natural forest structure.

With such changes afoot at home, Danish forestry aid started to reflect this shifting focus. Henrik Keiding, another Danish student of Syrach-Larsen and the first director of the Teak Improvement Center in Thailand, describes the change: "The main priority was originally given to relatively few species. Today, the work involves protection of a much greater number of species and ecosystems. In this way, a broader genetic base is secured and a greater number of possible options to respond to tree planting needs in the future... In other words, teak's 'status' in Thai-Danish forest genetic cooperation today has changed from being the dominant species to one of many valuable tropical species."[43]

Danish forestry aid also changed its political character. From its beginnings in 1965, aid to Thai teak research was purely a "technology transfer" project. Three decades on, the aim of the work, according to Erik Kjær of the Danida Forest Seed Centre, had turned to supporting poor people. "That's why we're not involved in big plantation programs," he says. "It was a shift from pure plantations and looking at trees as factories producing a single commodity, to people and how they use the forest. Big companies don't need that kind of support. They can manage on their own."[44] In recent years, the focus has expanded to cover other so-called "multi-purpose" species more useful to local communities like acacia and neem, he says.[45]

Through the 1990s, this change could be seen in the Danish financing of a major program in the Thai forestry department called the "Forest Genetic Conservation Management Program." Its focus spread far beyond *Tectona grandis* to include seed production of dozens of tree species (though still no non-timber forest products). It was run by Jens Granhof, a Danish forester who has worked in Thailand for thirty years at both the Teak and Pine Improvement centers. The Danes also financed research and projects related to the relationship between forests and people.

While they welcomed the broadened focus on a larger variety of tree species, Thai grassroots NGOs have accused Denmark of being oblivious to the political landscape, which is characterized by conflicts between Thai lowlanders and ethnic non-Tai highlanders over land and water, and growing pressure from the forestry establishment and conservation groups to resettle forest-dependent communities. As anthropologist Pinkaew Laungaramsri writes, the Danes' "analysis prevents [them] from understanding the root causes of environmental degradation—the conflicting approaches between Western-based 'scientific' forest management enforced by state foresters, and local systems of natural resource management."[46]

Teak: Green Revolution Forestry

Teak is not only one of the most important timber trees of tropical countries, but it is at the same time well suited to be planted for the formation of even-aged and uniform plantations.

Carl Syrach-Larsen, Danish forest geneticist[47]

In the front hallway of the Forestry Industry Organization's office in Bangkok stands a sad tribute to the lost glory of Thailand's teak forests: a massive teak stump measuring almost 2 m across. A cursory counting of its rings puts its age at about three hundred years. No one knows how old a teak tree can become, though there stands a giant teak tree in Uttaradit Province that is estimated to be a thousand years old. Its girth is almost 10 m.[48] It is known, however, that teak takes about 150 years to reach a girth of over 200 cm, which was the minimum allowable size of trees harvested in Siam until the 1930s.

By the Second World War, the average size of logs coming out of the Thai forests started to decline, and with this an intensification of the interest in the establishment of teak plantations to replace the depleted natural supply. In Thai forestry discourse, tree plantations are depicted as an integral part of sustainable yield practices, an extension of the basic principles of scientific forestry. In fact, teak plantations were pursued as a result of over-cutting; that is to say, they are the consequence of the failure to implement sustainable forestry as it was meant to be practiced.

In some ways, Thai foresters were lucky with teak. As the Danish forest botanist Carl Syrach-Larsen observed, teak lent itself to planting on bare land for three reasons:

1. teak is a pioneer species,
2. it is light-demanding, and
3. it is relatively fire-resistant.

Incidentally, these three characteristics of teak might also help to explain why colonial foresters noticed a predominance of young teak trees growing in the fallow fields of swidden cultivators in the border areas between Burma and Siam.[49] But the purpose of the modern teak plantations was not merely to compensate for the depletion of natural teak. It also aimed at shortening the time required for producing timber, improving and standardizing its quality, and increasing its commercial volume. This was Green Revolution thinking applied to forestry.

In Scandinavia, Syrach-Larsen was known as a "forest wizard" and a "man of ideas." Danish foresters saw in him the possibility for the country to "send skilled foresters out into the world and... contribute to Progress through our knowledge."[50] The British forester H. G. Champion spared no words in his praise for the Dane: "With the importance placed on maximizing production it is not surprising that there is considerable interest in obtaining the best possible material for plantations. In Dr. Syrach-Larsen, Denmark has one of the best men in the field of tree breeding and it is thanks to his skill and enthusiasm that such significant progress has already been made..."[51] Syrach-Larsen himself admitted that "we forest people are a bit like missionaries."[52]

Syrach-Larsen was instrumental in the first botanical survey of Thailand financed by the Danish government. When he arrived in Lampang in the mid-1950s as part of the Danish botanical expedition, he brought with him brand new ideas about forest genetics that would for the first time make this possible. He believed that "forestry in its silvicultural aspect must turn its attention to the use of genetics" because while the great old trees of the forest may be more pleasing to the eye, "we must bear in mind that they are past their most favorable economic age and that greater benefit is obtained from the less attractive, but more uniform forest managed commercially." Forests should be improved in the following way: "The inherent tendencies (genes), which we have found in nature and, by our skill, have caused to lodge in the seed, will continue to work for us right up to the felling of the tree. An innate capacity to withstand an adverse climate or to overcome disease, a tendency to have more slender branches which fall off more readily and give more knot-free timber, or an innate tendency for vigorous growth can be fixed in the seed and remains as an inherent tendency throughout its whole life. If we are to plant and grow forests, it is thus of the greatest economic importance to seek out and utilize such tendencies."[53]

Syrach-Larsen took inspiration from industrial agriculture. He was impressed by the massive gains in productivity exhibited by American hybrid corn and noted that "forestry work in this field is of no account in comparison with the marked advance which has been made in agriculture and horticulture." Foresters had watched the progress being made in crop genetics "with envy" for too long. But simply adopting the techniques of agricultural researchers was not an option for foresters. Trees are tall in comparison to wheat, for example, and it is difficult to get at the flowers for the purpose of controlling pollination. The most desirable trees can be dispersed over long distances, making breeding cumbersome. Moreover, while agricultural breeders can take advantage of rapid succession of generations, ten, twenty or even fifty years must pass before a tree will flower and set seed. For Syrach-Larsen this made matters more complicated, but not necessarily insurmountable, for foresters. "*We should go our own way*," he urged.[54]

Nowhere in the tropics did Syrach-Larsen have a better opportunity to test his ideas than in Thailand, where his own connections with the East Asiatic Company and Denmark's strong historical ties with the country must have helped open doors.[55] While his contemporaries tended to view the forest crudely as a network of compartments to be felled and then replanted, Syrach-Larsen looked at Thai teak forests in a totally new way. He saw them as groups of individuals with variations in branching formations, flowering rhythms, crown shapes, and lengths of stem.[56] He hypothesized that if particular traits of teak individuals could be manipulated and optimized, the timber-producing capacity of trees could be greatly improved. He wrote in 1966: "We were rather intrigued to notice the diversity to be found in these populations, from trees with a fine clean trunk of considerable height to trees with a very short stem topped by a shrub-like crown ... *The genetic variation in age of individual trees at the onset of their first flowering phase, and thereby the length of the straight axis above ground, is of major technical importance.*"[57] Syrach-Larsen observed that early flowering resulted in bad stem formation, and therefore hoped that this could be improved through breeding. After the first five years, he concluded that five of their clones preserved flowering characteristics over the period. This suggested that flowering was genetically determined, and could be further influenced through selection and breeding.[58]

Syrach-Larsen's approach inspired a generation of Thai forestry researchers to pursue an ever-narrowing genetic path. Prominent among them is Thai forest geneticist Apichart Kaosa-ard, who has worked with the Thai teak research program since the 1970s and is now one of the world's foremost teak experts. Apichart says Thai teak research has since Syrach-Larsen's time been guided by three main concerns: accelerated growth rate, optimum stem form, and pest control.[59] In other words, the idea had been to reduce the necessary rotation period by selecting individual "plus trees" that had a higher than average diameter/height growth rate. Foresters were also after a clear bole (or trunk) with no defects or knots. The final consideration, given the special vulnerability of pure plantations, was resistance to pests and disease. "If we delete the external factors with the same spacing and same silvicultural management, the remaining variation must be due to internal (genetic) factors," explains Apichart.

Based on these criteria, researchers selected individuals to be used as "mother trees." Clone banks were established starting in 1965 and tests were performed on individuals. Three decades of research produced five hundred clones from 2,000 hectares of clonal seed orchards at various locations throughout the country. Thai and foreign seed sources have been compared in provenance trials, budding and grafting techniques have been developed for clonal propagation of mature trees, and reproductive biology has been studied intensively.[60] The bottom line is an improved planting stock that in theory produces up to 25 percent more timber than its natural counterparts. That is, under the best conditions, trees that used to take sixty years to reach a girth of 150 cm now do it in forty-five years.[61]

Ironically, while this research has been applied in teak farms worldwide through the efforts of the Danida Forest Seed Centre, the results in Thailand leave much to be desired. As Erik Kjær puts it: "There is a lot to be learned worldwide from the mistakes made in Thailand." Such as they are, the teak plantations established by the Royal Forestry Department and the Forestry Industry Organization[62] have had limited success because they were often located on unsuitable soils and have, in any case, made little use of the improved seeds. A large part of the problem was that the villagers who were expected to plant and tend the trees often lost farm and forest land as a result and were offered pittance in return. Another difficulty has been inadequate seed supply. In 1986 the Teak Improvement Center was producing only 5 percent of the required seed with the other 95 percent coming from natural forests.[63] Moreover, Thai teak plantations have suffered the usual problems of single-species stands. With a build-up of leaves and weeds, young plantations have been particularly susceptible to fires, whether these were natural, set to clear agricultural or grazing land, or acts of incendiarism by resentful villagers. And resistance to the stem borer insect (*Xyleutes ceramicus*), the main pest in monoculture teak stands, remains a mystery in spite of the best efforts of researchers. Apichart suspects the increased incidence of stem borer damage is "probably due to the ecological change from the mixed species teak forests into the pure teak plantations." He suggests that mixing a few tree species could eliminate the problem.[64] But this has never been seriously explored by Thai foresters because, as Apichart puts it: "We have to make a distinction between conservation forest and production forest. Practically speaking, it has to be monoculture because managing is easier and investments are cheaper," he says.[65]

Researchers propose to solve the problem of inadequate seed supply in the future by increasing the use of cuttings from juvenile seedlings and tissue-culture plantlets. So-called "serial cuttings" (cuttings from plants derived from cuttings) can result in thirty to forty plants produced from one seedling in a single growing season. It is thereby estimated that 1 kg of improved seed can produce five to six thousand plants,[66] as compared with only ninety-two seedlings per kilogram using traditional planting methods.[67] One forester describes cuttings as "the final end of the breeding process... You want to find the optimal genetic combination of parents, then you select the optimum offspring and multiply them [through cuttings] *en masse*." The result is a plantation of near-total uniformity. The advantage is that individuals in the stand express only those characteristics deemed desirable by the breeder. The disadvantage is that any weakness in the cutting is multiplied over the entire area of the plantation, thus greatly magnifying its vulnerability. In other words, you "run the risk of an epidemic suddenly exploding."[68]

Because of such vulnerability, a natural teak forest of the type observed in Ngao by Syrach-Larsen forty years ago is needed as a constant source of new genetic material. Indeed, Syrach-Larsen warned of the crucial importance of maintaining natural diversity, otherwise "valuable inherent tendencies (genes) may be swept away by irrational felling of remnants of natural forests."[69] Ngao was the obvious place to establish the Teak Improvement Center in 1965 because the best Thai teak forests were at that time centered in this part of Lampang Province. Sadly, Ngao has been emptied of much of its teak. Today, most of Thailand's teak diversity remains within the boundaries of the Mae Yom National Park, whose survival is itself threatened by the planned Kaeng Suea Ten water diversion project.[70]

Niels Elers Koch is the current director of the Danish Centre for Forest and Landscape Planning, and an advocate of the "soft" approach to forestry in Denmark based on mixed stands of native species. Reflecting on Syrach-Larsen's influence on Thai forestry thinking, Koch says: "From today's perspective, he used industrial agriculture's methods in forestry."[71] But Koch warns against judging yesterday's thinkers by the standards of today and insists that the old Danish geneticist was a broad-minded thinker with a vast range of interests. "He was not a production-oriented man. He was interested in forest history, in conservation. He was a gentle person. He was not a timber beast."[72]

Pine: A Window on Colonial Forestry

> *You say you want to help people to have a better quality of life. It's as if you're taking away my elephant and giving me a chicken in return ... You tell us that if you leave [the forest] alone, it will become degraded, but if you log it, you can save it. What kind of science is this?*
>
> Karen elder, Ban Wat Chan, Chiang Mai[73]

By force of ecological circumstance, Ban Wat Chan in the Mae Chaem District of Chiang Mai Province has never been logged on a large scale. Mae Chaem River is the main tributary of the Ping, one of four rivers flowing southward feeding into the Chao Phraya. Though it lies in the Thai timber heartland, the area around Wat Chan was spared because throughout the forests of the sub-district the dominant tree species is not teak, but pine.[74] These forests have a markedly different feel from the mixed deciduous and dry dipterocarp forests that cover much of the surrounding areas. Absent are the complex layers of vegetation. Places around Ban Wat Chan are reminiscent of "Chipko" forests in northern India, or even of Canada and Scandinavia, with a slippery layer of needles carpeting the forest floor and wide-open spaces between the trees. Thai forestry professor, Somsak Sukawong, has hypothesized that a great fire in the past might explain these extensive tracts of pine.

During the course of the century, the adjacent forests were gradually stripped of their largest teak and other hardwood trees—first in 1913 by the Forestry Department,[75] then in 1927 by the French East Asiatic Company,[76] and after the Second World War by the Chang Pheuk (White Elephant) provincial timber company.[77] Today the catchment area contains Thailand's largest remaining natural pine forest. When plans were drawn up to log the pines of Ban Wat Chan in the mid-1980s, they marked the first concerted effort to extract timber from this area.

Prior to this, there had been a number of attempts to develop the use of pine in Thailand. Two decades earlier, FAO undertook a major study, the so-called Pulp and Paper Material Survey Project, whose purpose was to survey the extent of pine resources in northern Thailand—much as Herbert Slade had done with teak—and to investigate the possibility of developing a pulp and paper industry in Thailand. FAO's idea had been to establish a long-fiber pulp mill based on plantations of the two indigenous pines, *Pinus merkusii* and *Pinus kesiya*.

In 1963 a 60,000-hectare site at Baw Luang in the Hot District of Chiang Mai was chosen to set up pine plantations. The area to be used for planting trees appeared to be abandoned but, in fact, consisted of the fallow fields of Lua swidden farmers. FAO's plan was to hire local people to tend the trees instead of tending their farms, for which they would receive a daily wage. A contemporary report notes that "in contrast with the *taungya* system, the land in the pine plantation is not used agriculturally. Thus clearing and weeding are done by wage labor."[78] The Lua had been self-sufficient subsistence farmers and reportedly objected to their fallow land being used by the project. The report states that the "loss of subsistence products is compensated for by wages, but the net value of this form of labor, and the effects of transformation of these people from subsistence farmers to wage workers have not yet been studied."[79]

The FAO survey had recommended improving pine seed, along the lines of the work being done on teak, for the purpose of establishing pine plantations.[80] As the Danes were already funding teak research in Lampang, it was natural that their work be expanded to cover pine. The Danish-financed Pine Improvement Center was thus set up in 1970, and trials of both indigenous and non-native species of pine were carried out.[81] An FAO consultant's report declared that there was "no doubt that pines in the north offer great attraction for pulp and paper development."[82] The purpose of the center's breeding efforts was "to expand the pulp and paper industry in this country." Thai forestry professor, Lert Chuntanaparp, who participated in species trials wrote that "it is the prime object of this investigation to find the most suitable provenances to be used in the forestation program in Thailand."[83] Initial research attempted to evaluate the relative performance of various provenances with respect to their height, diameter, volume production, stem and crown form, branching habits, resistance to pests, and wood quality.[84]

Research at the Pine Improvement Center determined that pine tended to grow best in the north of the country, and that of the two local species, *P. kesiya* was preferable for growing in short rotations of between fifteen and twenty years. Years of experimentation with selective breeding have resulted in pine seeds that have a growth rate 25 percent higher than trees that grow in nature. A former director of the center, the Dane Jens Granhof, notes that this improvement could easily be doubled by increasing the intensity of selection and plantation management. As for pests, there is a much lower risk than, for example, with *Eucalyptus camaldulensis*, since these pines are indigenous to the region.[85]

FAO's pulp mill did not however materialize. The plan had been to build it in the north of the country near the new Bhumibol Hydroelectric Dam (1964) and to use the large reservoir both for transporting the logs and for dumping wastewater from the mill. To its credit, the Electricity Generating Authority of Thailand (EGAT) was worried that pollution in the reservoir would be carried downstream towards Bangkok, and it vetoed the project.[86] And since the Gulf of Thailand is several hundred kilometers to the south, the prospect of setting up a pulp mill on the coast fed by pine wood growing in the north proved economically unviable.

But there were other problems. Whereas pine has been a staple of the European and North American wood industries, logging pine in Thailand on an industrial scale proved difficult. Unlike hardy teak, for which many years could elapse between girdling in the forest and eventual delivery (typically, 1,000 km downstream), newly felled pine is susceptible to insects and to blue stain fungus, which discolors the wood.

The Royal Forestry Department put considerable efforts into promoting the species.[87] But Thai timber companies showed little interest. Reungchai Pao-sujja, who took over from Jens Granhof as director of the Pine Improvement Center on completion of his doctoral studies in forest genetics, described it this way: "They tried to cut it, but all the wood would be destroyed by fungus and the quality was just not good enough ... In the end, it wasn't worth it."[88]

As for the Forestry Department's pine plantations at Baw Luang, some 16,000 hectares of *Pinus kesiya* were planted between 1965 and the 1980s. But these did not produce the expected yields. Since the logging ban, the trees have not been thinned and have therefore become scrawny and too tightly spaced. In Granhof's opinion, the best that can be done is to cut them all down and start over.[89]

The failure at Baw Luang and the industry's lack of interest in the species may help to explain the new approach taken by the Forestry Industry Organization in the early 1980s. FIO would bypass the time-consuming task of establishing a plantation from scratch by targeting the full-grown timber in the natural pine forests of Ban Wat Chan. This time, aid funds would come not from Danish foresters and the FAO, but from Finnish foresters and the Nordic Investment Bank (NIB). With their internationally-renowned expertise in the felling, planting, and processing of long-fibered timber, the giant Helsinki-based forestry consultancy firm Jaakko Pöyry Oy International proposed to use what they called "hot logging" to overcome the difficulties in Thailand. This meant removing logs from the forest immediately after they were cut, treating them with chemicals to prevent blue stain fungus infection, and processing them at a sawmill in the middle of the forest.

Jaakko Pöyry's plan for the forests of Ban Wat Chan[90]—dating not from 1885 but 1985—resembles a classical colonial forester's approach where a clean matrix designed to maximize the planners' profits is imposed on messy reality, and people are simply expected to get out of the way. The tactics that the Finnish consultants initially used to try to convince villagers to give up their customary rights to forest land reveal once again the dual nature of scientific forestry—a project of biological simplification, on the one hand, and the imposition of political control over land, on the other. The plan was to log 4,000 rai (640 hectares) of pine forest each year over a twenty-five-year cycle, 2,000 rai by the FIO and 2,000 by local people. Each of the four hundred families was to be assigned 125 rai of land for which they were responsible. They were expected to abandon agriculture. Instead, they would clear cut 5 rai each year, plant new saplings and protect their area from fire. In the following year, they would move on to the next 5-rai plot. By the twenty-sixth year, they were to return to the original plot, which was by then to have reached harvestable size, and start the process again. They were to tend the forests instead of growing rice, and to buy food with money they received for fire protection duties and as stumpage fees for timber sold to the project's sawmill. Their wage was set at 40 baht per day.

The inhabitants of Wat Chan Sub-district are Karen, several hundred families in seventeen villages who have lived in these pine forests for at least two hundred years, according to local elders. They say they were granted rights to the forest lands of the Mae Chaem watershed by the *chao muang* of Chiang Mai, their title written on a piece of tiger's hide. Two centuries later, they still claim rights to the land, though they have no modern, legally recognized title.

Local people first became aware of FIO's plans when roads started being built into the area, and some timber was taken out of the forest. One elder described the general confusion: "They just cleared land all over the place. They built roads right across our fields and messed up the *mueang fai* channels... People were very angry. We couldn't believe they were taking out the trees. We had several meetings to complain, and the FIO officials threatened to send us back to Burma."[91]

By 1984 people in Ban Wat Chan had become familiar with the name "Cha-ko" (Jaakko Pöyry). The deputy head of FIO's Ban Wat Chan project, Suntisuk Prasitsak, saw Cha-ko's intervention as the only way to guarantee the forest's survival. In what sounds today like a direct quote from a nineteenth-century forestry textbook, he declared: "If you leave it to nature, the forest will die off and eventually be completely gone... the old trees are being wasted, just standing there without being used."

The Karen were not only fearful for their future. They were also doubtful of the project's viability. At a gathering in 1992, village elders offered a critique of the foresters' plan: "It's impossible to calculate nature. An old pine tree has so many seeds, but how many actually become big trees?... They can't just use those theories of forestry here. The spirit of the people is different, so is the land. If these big old pine trees are gone, they won't ever come back... We think of future generations. If there is no more forest, where are they going to live?" Another worry was the risk of tampering with an already precarious environment. "You have to be especially careful with pine forest. It isn't like dry evergreen or deciduous. The soil is already very dry and acid—hard to work."

In addition, the community leaders were wary of the Finns' emphasis on protecting against fires at all times. Echoing a debate that has been a source of contention among foresters since the nineteenth century, the Karen said that their method of controlled annual fires did not damage large trees, and at the same time prevented a build-up of dry needles. They warned that if the pines were left unburnt for too long, and a big fire took hold, the whole forest could be lost.

More than fire or the questionable viability of young pine farms, however, the Karen were concerned about how the logging project would affect their water supply, agriculture, and forests. An elder described the community's dependence on an intact ecosystem: "If they remove the trees from the watershed, the rivers will run dry, the soil will lose its fertility, and we won't be able to grow rice. How will we eat?"

The 1985 plan was impossible to push through. In a climate of growing environmental awareness, there was local opposition, though at first local people were afraid to express it openly.[92] As protests grew, even FIO director M.R. Adulyadet Chakraphand admitted in an interview that Jaakko Pöyry's plan was "just a clear-cutting scheme."[93] After the 1989 logging ban, the idea of logging Thailand's last stands of natural pine became even harder to justify. But as the country's main concession holder, the FIO found itself in a desperate financial situation when the timber licenses were canceled. An obscure cabinet resolution exempted the forests of Ban Wat Chan from the ban, however, and this provided the necessary legal loophole for FIO to move ahead with its plans.

Thus in 1991 FIO again hired Jaakko Pöyry, which was by then deeply involved in writing a "master plan" for Thai forests that would include a new proposal for Ban Wat Chan. This time, they were to take the environment and local people's general health and welfare into account. Their "clear-cutting scheme" was thus transformed into an "integrated rural development" plan for Ban Wat Chan.[94] "Development," in this case, involved tacking on a health clinic, a school, and some agricultural extensions to the original logging scheme. Predictably, the plan still blamed local people for destroying forests. The purpose of the project was now to stop environmental degradation, to achieve food self-sufficiency on a sustainable basis for the Karen people, and improve their land use through watershed and forestry management. This would not only "stop the deforestation," it would conveniently "save forests also for potential commercial management" as "most of the forests would be ready for cutting now."[95]

Jaakko Pöyry's consultants calculated that the average Karen family's annual income was about 3,000 baht. If each family sent one member to work with the project, they would quadruple their income, earning up to 12,000 baht a year.[96] While the form was reminiscent of traditional *taungya* systems, this wage was significantly higher than salaries paid to members of FIO forest villages.[97]

Forestry officers in British Burma had tried in similar ways to bribe the Karen away from their life in the forests, and encountered widespread resistance. A forester's report from 1864 reflects a very similar conflict: "They have no desire to move from tracts where their forefathers have lived . . . Several [headmen] admitted that the remuneration for sowing teak would be ample to enable them to buy a pair or two of buffaloes in the course of two or three years, but that at the same time they had no wish to do so."[98]

The Finns were clearly aware that they would have to deal with opposition from local people, the ostensible beneficiaries of their development project. As part of the process of writing the new plan, a company consultant visited Ban Wat Chan in 1991 to assess the level of resistance. According to the Karen that were interviewed, the consultant asked village leaders why they were opposed to the logging, and offered them three options for dividing the timber income. The first was by family, as the 1985 report had suggested. The second involved groups of farmers working in a given catchment area. They would log the area collectively and be given a lump sum to be divided amongst them. The final scenario was for whole villages, or groups of villages, to be responsible for fire protection, cutting, and planting in larger areas.

The Karen told the consultant they did not want any logging in the Ban Wat Chan area. One village leader offered his analysis of the Finn's three offers:

The first, separation by family, struck him as a shameless "divide and conquer" strategy that would very quickly destroy cohesiveness in the community and, likely, the forest along with it. The third, division into groups of villages, risked that individual leaders could be bought off to make bad decisions on behalf of the majority. When pressed, he acknowledged that if forced to choose, he preferred the second option because it would give the community the greatest control over which parts of the hillsides would be cut to cause the least possible damage. Even if some of these groups were willing to sell out, most people, he felt sure, would prefer to limit the cutting to the least harmful areas.

Local protests continued through 1992 and 1993. They included several tree ordinations, in which Buddhist monks, animist shamans, and villagers held ceremonies to bless the trees in the forest. Monks' robes were even wrapped around the sawmill in an attempt to discourage FIO's workers from participating in the logging scheme.

The Ban Wat Chan logging project was eventually canceled, and the sawmill was dismantled in 1994.[99] Having failed twice to extract the timber from the area's forests, FIO subsequently turned to promoting Ban Wat Chan as an eco-tourism destination where healthy pine forests are a main attraction.

Eucalyptus: Notorious *camaldulensis*

> *The gum tree has a pale ragged beauty. A single specimen can dominate an entire Australian hill. It's an egotistical tree; standing apart it draws attention to itself and soaks up moisture and all signs of life, such as harmless weeds and grass, for a radius beyond its roots, at the same time giving precious little in the way of shade.*
>
> Murray Bail, *Eucalyptus: A Novel*[100]

Native to Australia with more than five hundred species in the genus, eucalypts are magnificent trees. The characteristic drooping leaves of the mature *Eucalyptus globulus*, one of the most widely planted of the species, emit a minty scent and the bark peels in paper-thin shreds, covering the massive trunk with layers of pink, cream, pale green, mauve, and brown. Eucalypts have, however, attracted the attention of foresters not because of their beauty or their fragrance, but because of their uncanny capacity to grow fast. A Finnish forestry professor, speaking at a tropical forestry-training course in Helsinki, described its importance: "The eucalypts produce some of the heaviest, hardest, and most durable woods known. The quality of the timber, coupled with the rapid growth rate, regenerative powers, and often great size, makes this genus the most valuable source of hardwood in the world."[101]

Of all the eucalypts, *E. camaldulensis* has the broadest natural ranges, occurring in all Australian states except Tasmania.[102] *Camaldulensis* is also deemed the most suited to climatic conditions in Thailand, and has become the favored tree species for producing pulp for the paper industry.[103] It can grow in almost any kind of soil, whether drought-prone, waterlogged, saline, or acid.[104] While there is much variation within the genus, eucalypts in general grow fast because their extensive and deep root systems allow them to use all available water. Eucalyptus is therefore both well adapted to dry conditions and grows exceedingly fast in areas with a plentiful water supply.[105]

Moreover, while individual trees may grow faster and bigger if planted at greater distances from each other, the total per rai volume increases when trees are planted tightly together.[106] One study showed that trees planted with wide spacing reached almost double the diameter of those planted tightly together (see table 2). However, the results for volume production were the opposite. Trees planted close together produced almost double the volume of timber per rai (17.4 cu m per rai) of those planted at 6 x 6 m spacing (just 9.9 cu m per rai). Studies like this one made it clear that for use as raw material to produce wood chips for pulp, the optimum-planting regime for eucalyptus in Thailand was tightly spaced, intensive, large-scale, monoculture stands.

Table 2 Growth and wood production of four-year eucalyptus at various planting distances

Planting distance (m)	2 x 4	4 x 4	4 x 6	6 x 6
Diameter at breast height (m)	12.72	15.49	17.56	19.24
Dry weight of bole (kg/tree)	49.233	75.606	99.988	121.017
Volume (cu m/tree)	0.091	0.140	0.185	0.224
Total volume (cu m/rai)	17.443	13.485	11.66	9.964

Source: Pongsak Sahunalu, et al., 1987

The eucalyptus genus is for today's tropical foresters what teak was to their colonial forebears. All Western donors within the Thai forestry sector have promoted the planting of eucalypts. With natural forest loss continuing relentlessly, teak planting advancing at a snail's pace, and unhappy prospects for pine, Thai foresters also embraced *Eucalyptus camaldulensis* as the embodiment of a new hope for the forests. And, until the conservation agenda came to dominate, this new hope extended to forestry as well, as a chance for the Royal Forestry Department to reverse the dismal deforestation statistics and, once again, earn money for the state. The near unanimity and enthusiasm with which Thai foresters stuck by the claim that the species is beneficial both to the environment and to small farmers says volumes more about the forestry discipline and bureaucracy than it does about the tree.

❖ ❖ ❖

The first recorded eucalyptus tree was planted in Phrae in 1946 by Sa-ard Boonkird who joined the Danish botanical survey a few years later.[108] In 1950 the Royal Forestry Department made experimental plantings of five species of eucalypts—*E. alba, E. saligna, E. paniculata, E. pitriodora*, and *E. grandis*—at the Doi Suthep Forestry Station in Chiang Mai. Then in 1954, as part of FAO's pulp and paper survey, the seeds of fifteen eucalypt species were imported from Australia and planted in provinces representing the four regions of Thailand: Chiang Mai, Si Sa Ket, Kanchanaburi, and Surat Thani.[109] The first *camaldulensis* species of eucalyptus was planted the following year at the Thai-Danish Pine Improvement Center in Hot. Reungchai Pao-sujja initiated trials of several eucalyptus species in 1973 and found *camaldulensis* to be "clearly superior."[110] The following year, the Danish aid agency DANIDA, which was funding Reungchai's work, supported the establishment of an FAO/UNEP program for *ex situ* gene conservation of Central American pine and eucalyptus species.[111] More eucalyptus trials were funded by the UN Development Programme (UNDP) in Surin Province in 1977.[112] Four years later, the Japanese International Cooperation Agency also established trial plantations in the northeast.[113]

According to an outspoken proponent of eucalyptus, forestry professor Boonwong Thaiutsa, the real awareness about planting eucalyptus trees in Thailand started as a result of the Forestry Department's agro-forestry research at Huai Tha in Si Sa Ket in 1978.[114] This twelve-year research project headed by Pitaya Petmak of the department's Forestry Research Division sought to compare the effects of *Eucalyptus camaldulensis* and three other tree species on the yields of various crops. Citing Pitaya's research, Boonwong asserted that crops planted with *E. camaldulensis* had "high average yields." It was this research, he claimed, that prompted the Forestry Department to initiate in 1981 the widespread promotion of eucalyptus planting by villagers in the northeast for use as fuel wood. Later, companies like Shell Thailand and the Suan Kitti Reforestation Company would also refer to Pitaya's research as proof of the environmentally beneficial effects of eucalyptus. As we shall see, however, Pitaya's own assessment of his results differs radically from those of his supporters.

The planting of eucalypts met with local resistance from the very beginning. From Surin in 1977, where indignant farmers tore UNDP's saplings out of land they used for cattle grazing,[115] farmers throughout the northeast and east began to complain about the impacts of the eucalyptus plantations. They cited water shortages, soil degradation, reductions in the productivity of adjacent crops, replacement of existing natural forest, and conversion of rice paddies to tree farms.

Viset Sinet, a village leader in Yasothon Province, remarked that foresters encouraged them to plant crops "between the rows" of eucalyptus trees. "Though in the first year the problem doesn't show up much, later on, particularly in the fourth and fifth year when the eucalyptus trees have reached maturity, the land is too poor for other crops to survive. Moreover, eucalyptus roots leave no space for other plants' roots to take hold."[116] Sanitsuda Ekachai of the *Bangkok Post* spoke to Chand Poonsawad, a farmer and mother of nine in Chachoengsao Province, the heart of eucalyptus country. Chand explained that she and her husband sold their land to the Suan Kitti Reforestation Company to pay off farm debts incurred a decade earlier due to falling crop prices: "Since the plantations, the land has just dried up and the rice crops have failed. It was never like this. Water for the paddy fields must be pumped up from the canal, which becomes drier every year. But when the annual flood comes, water does not go away as quickly as before. The dikes around the plantations keep the fields flooded, destroying our rice crops. I'm a rice farmer, but now I have to buy rice to eat."[117]

In 1985 the area of eucalyptus planted by private companies reached almost 5,000 hectares, with the Forestry Industry Organization planting an even larger area.[118] That September, two thousand villagers in Si Sa Ket Province dug up eucalyptus saplings and burned nurseries to create pressure on the government to forbid further planting—a seminal act of what forest historians call "incendiarism." Farmers elsewhere held rallies, wrote letters, protested, burned down plantations and nurseries, and ripped out saplings. Thai foresters continued to insist that there was no scientific evidence to support the farmers' claims, either internationally or in Thailand. Instead, they argued with vigor in favor of *Eucalyptus camaldulensis*, willing, it often seemed, to overrule common sense in defense of what had become Thai forestry's sacred cow. Villagers' claims were dismissed and critical questions from non-foresters were labeled uninformed or politically motivated. Pointing to the lack of debate among foresters on the subject, Rapee Sakrit, former rector of Kasetsart University, commented that "there is a tendency in the state bureaucracy to establish its own kingdom ... eucalyptus seems to have grown up and set its roots deep in the forest department. Its poison may be more profound than toxic substances. It is a tool of an economic system which permits outsiders to encroach on the rights and freedoms of villagers who are equally eligible to a decent life."[119]

Full state backing from the forestry establishment was, however, not sufficient to guarantee the spread of eucalyptus. Protests continued unabated and, as a result, private companies were able to plant only half the area in 1986 of the previous year.[120]

To put an end to the criticism, the Royal Forestry Department in cooperation with the Royal Agricultural Science Association and the Soil and Fertilizer Association organized a seminar in 1987, entitled: "Is Eucalyptus Really Toxic?"[121] The three-day event produced the following statement: "Eucalyptus has high commercial potential both domestically and internationally. But the promotion of eucalyptus should be governed by a planting policy that is in accordance with market and other conditions to ensure that production and market demand are in balance ... The results of scientific research conducted by state agencies in Thailand conclude that eucalyptus trees are not toxic, and have no clear negative impacts."[122] Foresters declared that "the planting of large, monoculture stands will be studied carefully." To this day, with the exception of Pitaya Petmak's research, no in-depth, long-term study of the environmental impacts of large-scale eucalyptus plantations has been carried out by Thai foresters.

According to Professor Boonwong Thaiutsa, confusion about the species persisted.[123] By "confusion," he was referring to the public criticism of eucalyptus promotion that intensified following the arrests of the Suan Kitti plantation workers in January 1990. The Kasetsart University forestry faculty organized a follow-up seminar in April 1990 entitled "Are We Still Worrying About the Toxicity of Eucalyptus?" Professor Boonwong, who chaired the meeting, explained why it was necessary for the experts to speak out: "We realized that the people making noise about this issue have no training in forest, soil, or water management. But at the same time, we professors have been silent." For Boonwong, the two seminars brought "closure" to the eucalyptus debate. Persisting criticism was unfounded and eucalyptus was being "victimized."[124]

Notwithstanding the confidence with which Boonwong and his colleagues trumpeted the benefits of eucalyptus, many others had documented potentially harmful effects of large-scale eucalyptus planting that, at the very least, suggested that there were grounds for raising serious questions.

FAO, a long-time promoter of eucalyptus, published *The Ecological Effects of Eucalyptus* in 1985, summarizing the international literature on eucalyptus up to that time. Though its authors noted that it was difficult to generalize given the variation among species and diversity of local conditions where eucalypts have been planted, the report conceded that "there is evidence from the humid tropics that young, rapidly growing eucalypt plantations consume more water, and regulate flow less well, than natural forests."[125] Moreover, the strong surface roots of eucalypts, which can grow out to twenty meters, "compete vigorously with ground vegetation and with neighboring crops in situations where water is in short supply." Where the trees are cropped regularly, there is a "rapid depletion of the reserve of nutrients in the soil. This is a direct consequence of their rapid growth." FAO acknowledged that eucalyptus plantations "largely displace the ecosystems that were there before. The relative importance, both ecological and social, of these original ecosystems should be carefully balanced against the advantages to be gained from the new plantations."

Curiously, in Thailand it was not just the FAO report but the very study that foresters and companies cited most often as proof of the ecological benefits of eucalyptus that should have sounded the alarm bells. When I asked Pitaya Petmak in 1990 why so many people were using his work to defend their promotion of eucalyptus, he admitted that he was perplexed. "Why do they only refer to my project? It's only one small study, and the objective had nothing to do with large-scale plantations," he said.[126] He maintained that there existed no definitive information on the effects of large monoculture stands of eucalyptus, and wondered why the Royal Forestry Department or other agencies within the Ministry of Agriculture had not undertaken such a study. "[Eucalyptus]," he said, "is like fire. If you understand its nature, it can be extremely useful. But if you use it in the wrong way, it will cause untold damage."

As Pitaya described it, the purpose of his twelve-year study, which began in 1978 in Si Sa Ket Province, was to try to establish a working agro-forestry model for poor farmers in the northeast.[127] He wanted to determine which tree species were best to plant along with other crops on small plots of land, giving farmers the most wood in the shortest time with the least detriment to their crops and the surrounding environment. Working on the assumption that large plots of empty land are impossible to come by, and that most farmers have only ten to fifteen rai of land on which to work, he designed a twelve-year experiment to compare the relative effects of four types of trees: *Eucalyptus camaldulensis*, acacia (*krathin narong*), leucaena (*krathin yak*), and peltophorum (*nonsi*). "I knew that eucalyptus wouldn't improve the soil and might harm the environment ... I also knew that nitrogen-fixing trees like acacia had uneven branches that are less useful for villagers' daily needs," he said.

Pitaya set about testing the effect of each type of tree on crops like corn, peanuts, rice, and sorghum. On a 16-rai plot of land, he planted four separate plots, each containing the four different tree species (1 rai each) planted at a distance of 2 m with an 8 m row in between. In these open spaces, various cash crops were planted, and their productivity compared. The first two plots of trees were cut down after four years, which Pitaya determined to be the optimal age to harvest eucalyptus to get the highest wood yield with the least investment of time and energy, and to avoid serious soil degradation. The third was harvested in 1986, and the final plot in 1990.

One of Pitaya's findings was that crops could only be planted between the rows of saplings during the first and second years. After that, the trees became too big and affected crop productivity. In general, he concluded that eucalyptus could prove a valuable tree for villagers if they learned to use it correctly. Otherwise, the trees could create more problems than they solved.

The odd thing about eucalyptus is that while the species uses *less water* to produce a given amount of wood than almost any other tree, it ends up using *more water* than most trees because it grows so fast. Pitaya cited an example of villagers who had planted *Eucalyptus camaldulensis* in a circle around their water pond, and were dismayed to discover that the water dried up completely. "Obviously this would happen because eucalyptus uses more water than any other tree," he said. In another case, villagers complained to him that eucalyptus planted near some mango trees appeared to reduce the growth of the fruit. Again, he said that because of this special characteristic of *camaldulensis*, the tree must be planted at a sufficient distance from other crops.

Pitaya found that eucalyptus and acacia produced about the same amount of wood after four years, and allowed for similar crop yields during the first two years after planting. But a difference could be seen in the plot where the trees were left to grow for eight years. He used corn as an indicator species, measuring its yield as a reflection of soil quality. Corn planted in the eighth year after the harvest of acacia had almost twice the yield of the corn in the eucalyptus plot. Pitaya explained that the "return uptake ratio" (the nutrients that are eventually returned to the soil) is known to be much higher in acacia, but it takes between six and ten years for the soil to improve. Peanuts, on the other hand—which are thought to be sensitive to toxins present in eucalyptus leaves—had statistically equivalent yields in both plots, suggesting that the toxins evaporated or dissolved with rainwater, thereby not affecting the growth of surrounding plants.

"Toxins are not really the problem with eucalyptus... but after eight or twelve years, the soil is definitely degraded because this is not a soil-improving species like acacia," he said. Pitaya's gut feeling is that villagers should not plant eucalyptus for more than four years at a time, and for no more than three cycles altogether, in order to give the soil a chance to regenerate. If the trees are allowed to grow for longer than four years, the soil has no time to "rest" and the stumps become too large for villagers to remove them manually. "Stump removal is no problem for the private companies, but farmers cannot afford to hire tractors to dig up their land," he said.

Another difference between villagers' eucalyptus plots and large private tree farms is that companies plant in rows of 2 x 2 m, 2 x 3 m or, in the case of Suan Kitti, 1.5 x 3 m. "This is not enough for farmers to plant between the rows, even in the first two years," he said. He suggested that a possible compromise solution would be for private companies to plant wider rows, allowing farmers to rotate their crops as the generations of trees were planted and harvested.

Pitaya Petmak's research did not attempt to measure the ecological impacts of large-scale eucalyptus plantations in Thailand. But a number of other studies did address some aspects of the subject. One of these, dating from June 1988, was conducted in Rayong Province by the Forestry Department's Watershed Conservation Division.[128] It compared soil moisture in an abandoned grassland with a three-year-old eucalyptus plantation during the rainy season (August–November 1987). Researchers found that soil moisture in the grassland was 30 to 35 percent higher during this period than in the eucalyptus plantation. It was found that organic matter (decomposing leaves) as well as the trees' root system induced leaching in the soil. Clay from the topsoil then settled down to the lower layer of the roots, making it difficult for rainwater to penetrate through this layer and recharge the groundwater below. "Consequently," the report states, "most of the rain water is delayed in the root zone stratum and is [absorbed] by the eucalyptus tree."

Another study, conducted like Pitaya's work in Si Sa Ket Province, by the Ministry of Agriculture's Northeast Rain-fed Agriculture Development Project, concentrated on the effects of eucalyptus plantations on the productivity of crops in surrounding areas.[129] The report noted that in 1982 farmers began planting eucalyptus on their paddy bunds because the trees were hardy, able to thrive in both dry and waterlogged conditions, and had an erect canopy that minimized shading in their fields. But in 1986, farmers started noticing a "detrimental effect" on rice plants in the vicinity of three- or four-year-old trees, and by 1987 a "similar but larger detrimental effect" was observed in other field crops.

"Rice and kenaf growth were markedly reduced in the vicinity of the trees and the effect extended for a distance of up to 20 meters into the crop. Plants close to the eucalypts were stunted and failed to thrive. The addition of higher rates of fertilizer did not compensate for reduced growth ... and cutting the trees did not solve the problem," states the report. While the authors warned that their results could not be considered conclusive because their data had not been subjected to rigorous statistical analysis, "they do indicate the presence of a potentially serious problem."

Published in 1989, a third study looked at the effects of *Eucalyptus camaldulensis* in several provinces of the northeast.[130] Researchers examined the impact of eucalyptus on the water table, soil physiochemical properties, comparative leaf decomposition rates, and on the growth of nearby crops. Their conclusions were as follows:

> large eucalyptus plantations can deplete underground water sources. Within eucalyptus stands larger than 70 rai, the water table dropped at least 93 cm lower than the surrounding land, causing possible difficulties for other cash crops with shorter roots.
> eucalyptus leaves fall in greater numbers than other crops, and decompose more slowly than rice, straw, mango, or acacia leaves, with the residual dry weight of eucalyptus leaves six times that of acacia leaves after a seven-month period.
> toxins in eucalyptus leaves inhibit the growth of other crops. Germination of maize, sorghum, sesame,

soybean, mungbean, peanut, and leucaena decreased
significantly in soil mixed with eucalyptus leaves. "The
reduction of germination of these crops was worsened
at higher rates of addition ... Four to eight fold
reduction of shoot growth was recorded in soybean,
mungbean, peanut, maize, and sorghum and as much
as 150- and 800-fold decreases were recorded in
leucaena and stylo respectively."

though eucalyptus is an extremely efficient wood
producer, it uses a higher overall amount of water than
other crops. "Eucalyptus uses only 1.41 milliliters of
water per gram of biomass [wood, leaves, and roots],
while pine uses 8.87 milliliters ... But the total yearly
water intake of eucalyptus is 1,200 milliliters, which
compares with only 760 milliliters used by pine."

In spite of Pitaya Petmak's own conclusions about his study, and the findings of these three others, Thai foresters have not been prompted to conduct any further investigation. They have largely maintained that critics are misinformed, politically motivated, and lack the necessary education to have an opinion about forestry matters. Instead, they lent full support to the 1985 forestry policy that aimed to facilitate the spread of eucalyptus plantations in Thailand.

A "New" Policy

Always ... move production to sites where the cheap wood is growing.

Jaakko Pöyry, Finnish forestry industrialist[131]

Teak, whose importance as an export commodity had been declining, finally disappeared from the list of Thailand's major export commodities in 1985.[132] By symbolic coincidence that same year, the Thai government approved the National Forestry Policy, which laid the foundations for a new direction in Thai forestry based not on logging teak and other hardwoods in natural forests, but on industrial plantations of eucalyptus. Though the Royal Forestry Department no longer exists in its original form and foresters have lost their mandate to manage the country's forests, this 1985 document remains a key policy document on Thai state forestry to this day.

Just as European teak-logging firms forced radical changes in forest administration and policy in late nineteenth-century Siam, so Thailand's new policy was a clear response to growing demands by companies looking for land to plant trees. In the heyday of teak, timber companies had to go where the teak grew. Initially, logging was facilitated by colonial administrations in India and Burma. But the overexploitation and temporary closure of forests in Burma caused them to move production to Siam. A century on, with environmentalism in industrialized countries growing and limited production capacity in Scandinavia, the industry's focus had switched to eucalyptus. Thailand appeared to offer a number of advantages over regional competitors to the potential tree planter: its relative proximity to Japan, the main Pacific paper market, a suitable climate, good infrastructure, a friendly policy environment, and cheap land.[133] *Pulp and Paper International* extolled Thailand's "political stability, low labour costs and . . . warm welcome for foreign investors . . . [as being] an irresistible magnet for new investment capital [for pulp-related activities]."[134]

By the 1980s, pulp industrialists were beginning to realize that their future depended on shifting operations to warmer climes. The tropics had become highly attractive for an international pulp and paper industry that was hitherto concentrated in North America and Scandinavia. Changes in pulping technology made it possible to make paper from hardwoods like the fast-growing eucalyptus.[135] With a reduced supply of Pacific Northwest old-growth timber and insufficient raw material produced to feed pulp mills in the Nordics from existing plantations, the industry was searching for new sources of wood chips. The increasing influence of environmental movements opposed to intensive industrial forestry, from British Columbia to

Australia, heightened the uncertainty about future timber supplies. Moreover, and perhaps most crucially, investors were attracted by the warm weather. Instead of waiting a century for trees to grow in the temperate and boreal zones, trees in the tropics could be harvested after as few as five years. These factors led to the establishment of substantial plantations in countries like Chile, Brazil, and Indonesia.

In a 1987 essay on the global outlook for the industry, Jaakko Pöyry, the founder and part-owner of the Finnish forestry company of the same name, argued that the main lesson from history was to "move production" to sites where wood is cheap.[136] This was a somewhat different message than that of sustained-yield forestry, in which the same forest, properly managed, was supposed to be used forever. Pöyry pointed to the example of the United States where pulp production relocated several times, from original sites in New England, to Maine, Ohio, and Wisconsin, to the southern states, and finally, when these forests were also used up, to the Pacific Northwest.[137] In 1989, the year of Thailand's logging ban, Pöyry commented on a "tightening" in the East Asian fiber supply situation as the demand for raw material for pulp production grew in the region. "One obvious way of securing raw material in the future is to invest in fibre production in fast-growing plantations in the Pacific Rim . . . companies should actively participate in reforestation within the region," he advised pulp industrialists.[138] Obligingly, Thailand's new forestry policy attempted to create the conditions that would encourage just this.

The policy was written at the request of Democrat Party chief Bhichai Rattakul by Sanga Sabhasri, the late former dean of forestry at Kasetsart University. The policy was ostensibly "environmental" in spirit, aiming to reforest the country after decades of forest loss. Its goal was to increase forest cover to 40 percent of the country's surface area, 15 percent for conservation purposes, and 25 percent as "economic forests." This would be achieved through private reforestation. At the rate the state was planting trees, proponents of the new policy often glibly pointed out, it would take foresters until the twenty-second century to reach the 25 percent target![139] Thailand's forests were being degraded due to "shifting agriculture, forest fires, [and] forest clearing by the hill tribe minorities," the policy noted. Government tree planting could not keep apace of the destruction, and the private sector—ready with money, technology, and an insatiable market for eucalyptus chips—needed land to plant trees. With an official forest cover of 28 percent (144,000 sq km), the shortfall of 12 percent (60,800 sq km) would be built up inside "degraded" state forest land, the so-called National Forest Reserves. Companies were offered attractive investment conditions, including tax holidays and a low rental fee (less than US$2 per hectare per year[140]), to rent out these lands to establish tree farms. For proponents of the policy, it was a perfect marriage of interests out of which Thailand could only gain. It was hoped that the policy would trigger the conversion of millions of hectares to eucalyptus plantations.

The opinions about how much of the 60,800 sq km should be covered with eucalyptus varied considerably. One estimate by the Thailand Development Research Institute (TDRI) approached the question by calculating the expected increased demand for raw material in the Japanese, Taiwanese, and South Korean paper industries. They came up with the figure of 5,000–8,000 sq km (3–5 million rai).[141] A 1987 study by the Ministry of Agriculture concluded that there were some 29,000 sq km of land inside Forest Reserves that were so degraded that "nothing else [but eucalyptus] could grow."[142] An ardent supporter of private reforestation, former Forestry Director General Yookti Sarikabhooti raised the stakes even higher. Speaking to a parliamentary committee on forestry issues, he proposed that the 60,800 sq km could be split three ways: 32,000 sq km would be planted with eucalyptus by the private sector, 19,000 sq km would be the responsibility of the state, while the remaining 9,800 sq km would go to villagers for community forests (of eucalyptus).[143] This assumed that 16 hectares (100 rai) be allocated for each of Thailand's estimated sixty thousand villages. A few years later, Jaakko Pöyry's consultants came up with a figure of about 25,000 sq km, which, if designated for "intensive management" by private companies, would make Thailand self-sufficient in paper production.

Many of these estimates turned out to be overly optimistic. By 1987 the total area covered by eucalyptus was just over 640 sq km—448 sq km planted by the state and the rest by private firms concentrated in Chachoengsao and Korat provinces.[144] By early 1992 the head of the Royal Forestry Department's Private Reforestation Promotion Office, Reungchai Pao-sujja, estimated that about 1,600 sq km of eucalyptus had been planted by the state and private companies in Thailand.[145] Seven years on, the Thai Pulp and Paper Association estimated that this area had doubled. But Manote Winaipanich of the Plantation Promotion Division insisted that this was an exaggeration, saying the real figure was closer to 2,000 sq km.[146] In 2005 plantations researcher Keith Barney estimated that eucalyptus plantation areas had reached between 4,350 and 4,800 sq km, while others say these areas could already have exceeded 6,000 sq km.[147]

While this is a significant area, it falls short of the optimistic projections made a decade earlier. The main reasons eucalyptus tree farms did not spread in Thailand on the scale that was predicted in the 1980s—in spite of full state backing and considerable capital—were farmers' protests, combined with the democratic space afforded by an elected government and a free press. With an estimated 10 million people living without title inside Forest Reserve land,[148] resistance was hardly surprising. One Thai villager, speaking at a parliamentary hearing in 1989, put the problem simply: "If all our land turns into plantations, where are we to go?"

The 1985 forestry policy marked an important departure from the past. Thai forestry was no longer primarily concerned with logging in natural forests. Even without a logging ban, the writing was on the wall, and the focus had switched to the establishment of tree plantations on degraded forest land. But the policy embodied a logical extension of old ideas. It was, in fact, an expression of colonial-style forestry in a more concentrated, unambiguous form. On one side, the timber bias was absolute. Monoculture stands of an exotic fast-growing species were by foresters equated with reforestation and the ideal "forest" obtained would have a *diversity of one*; a more "normal" forest than Dietrich Brandis could ever have imagined. The policy took no heed of ecological concerns. Indeed, it facilitated further forest destruction. Monoculture tree plantations were to be established on "degraded forest," which was defined not in ecological terms but in terms of the absence of large timber. More specifically, a degraded area was said to have no more than twenty 2 m tall trees per rai, or no more than eight trees of 50–100 cm girth per rai.[149] The new policy thus tended to encourage a process of degradation, often with villagers paid by companies to remove timber from an area such that it became sufficiently "degraded" to meet the Forestry Department's requirements for reforestation. Then, once the area was labeled as degraded, it could legitimately be cleared of all remaining vegetation to make way for new tree saplings, thus reducing or even eliminating the possibility of natural regeneration.

On the other side, the policy, like its nineteenth-century predecessor, criminalized customary use of forests by local people. By failing to acknowledge the presence of millions of untenured people known to be living inside forest reserves, the 1985 forestry policy further weakened these already disenfranchised communities. It rendered them politically invisible, thereby undermining their demands for land title. More importantly, the policy implicitly sanctioned, even encouraged, private companies to do what was necessary to get rid of them. This effective transfer of responsibility from state foresters to private forestry companies for the removal of people would become much more explicit in the coming years. The US$2 rental fee was thus purely symbolic in comparison with the real cost of moving villagers from the land—either by paying them to move, intimidating them into leaving, or by convincing them to give up their land claims in exchange for jobs as plantation laborers.

The National Forestry Policy of 1985 both intensified and complicated land tenure conflicts. Industrial plantations of eucalyptus, albeit privately owned, offered a way for the forestry bureaucracy to regain a degree of control over Forest Reserve land. Of course, the department was officially already in charge of this category of land. But by inviting powerful private firms to rent it out, state agencies could use companies as a buffer against landless people. The firms required large expanses of empty land. Perhaps they could succeed in getting rid of villagers where foresters had not. Foresters apparently expected that since the companies were lessees and not owners, the ultimate control would remain with the state.

This assumption may have been naive. Researchers at TDRI were among those who questioned whether corporations would, when their leases expired, willingly hand back valuable tracts of land after having paid farmers to leave, invested in roads and infrastructure, and built up plantations.[150] As TDRI put it: There was a "certain inconsistency between the lessees' expectations for eventual ownership of leased/purchased land and [RFD's] expectation of repossession of encroached land by proxy."[151] There was, however, no doubt about who would benefit from the set up. According to TDRI's research, large companies would earn as much as small farmers would lose: "While large-scale planters and companies make a healthy profit of 450 to 540 baht per rai per year, small-scale planters suffer losses of 48 baht per rai per year on the average... Under the normal circumstances of small-scale farmers—especially those with no secure land title who have access to only non-institutional credit at an average interest rate of 36 percent per year—losses as high as 550 baht per rai are likely."[152] In fact, the gap would probably be even larger as many of the hidden costs, such as environmental impacts, damage to crops, and stump removal (estimated to cost 1,000 baht per rai), which would be necessary after a few rotations, were not included in TDRI's calculations.

From the beginning, implementation of the forestry policy assumed a repressive character. Emboldened by the new legislation, foresters had a new legitimacy to their claim that landless people were illegal encroachers with no rightful claims to the land. Initially, they planted eucalyptus where it was easiest, in people's paddy fields and upland farms where the land was already cleared, or in areas used for grazing. When farmers protested, tactics changed and secondary "degraded" forests were instead cleared away and replaced with new saplings. Villagers' primary concern was about losing access to land. But the effects of eucalyptus on soil and water also became apparent as trees were planted adjacent to cropland. The foresters' "degraded" forests, moreover, often represented a diverse source of food, medicinal plants, and other things vital to local livelihood. One study conducted in the northeast showed that forest food consumption saved households 2,946 baht per year (36 baht = US$1), which is almost as much as the average annual income in the region of 3,076 baht. Replacing such areas with eucalyptus represented a direct threat to food security, especially for the poorest communities.[153]

In September 1985 a conflict broke out around the Non Lan community forest of Siew District in Si Sa Ket Province. A local company planted eucalyptus trees on top of the villagers' community woods and grazing grounds. Though the inhabitants had no legal rights to the area in question, they were outraged by what they perceived to be a clear case of trespassing. More than two thousand villagers responded by uprooting the young plantations, and burning down government nurseries, as well as a tractor that was being used to clear away existing secondary forest.

From Non Lan, similar confrontations spread throughout the region in response to the relentless promotion of commercial tree farms. In February 1987 inhabitants of Ban Nam Kham, Tambon Non Sawan of Roi Et Province submitted a letter to provincial authorities asking for companies to stop cutting natural forest for eucalyptus plantations. In April of the same year, there were protests in Tambon Khampia in Ubon Ratchathani Province by people asking authorities to stop planting eucalyptus and give them land rights. More protests followed elsewhere in May, June, July, and October. In March 1988 two thousand people from Tambon Kokekwang in Nong Khai, receiving no response from authorities to their pleas, cut down four hundred eucalyptus trees and burned saplings in a nursery. In June, three thousand villagers in Prachinburi burned down the houses of forestry officials after foresters had prepared the ground for eucalyptus planting.[154]

Speaking on condition of anonymity at the time, a Thai forestry consultant described the difference between the situations of Thailand in the 1980s and that of Brazil. "Reforestation has been successful in Brazil because villagers are more 'obedient' there. But in Thailand, even though villagers are living illegally in government land, they refuse to leave."[155] Thailand's problem, it seemed, was a shortage of repression. With marches, demonstrations, and petitions, and sometimes resorting to incendiarism and violence, villagers in many parts of the east and northeast confronted powerful state and private interests. "It was very tough fighting against [the government officials] because they have the law on their side and they see us as illegal encroachers," said a village leader. "But our demands are just. All we want is to be allowed to do the planting with species of trees that we choose ourselves."[156]

The most violent expression of the 1985 forestry policy was the military resettlement scheme, the so-called *Khor Chor Kor* of 1991, which foresters helped to plan.[157] When fully implemented, some 6 million people were to be forcibly resettled from their farms to free up land for tree plantations. Countrywide, people inhabiting 22,880 sq km of agricultural land would be squeezed onto 7,680 sq km, making available about 14,400 sq km for "reforestation."[158]

The forestry policy was actively pursued for about six years. With the cancellation of *Khor Chor Kor* in 1992 following massive protests, public support for private reforestation diminished and in the following year, the interim government of Anand Panyarachun passed a cabinet resolution that blocked major industrial tree farm expansion. The September 8, 1992, ruling revoked the policy of renting out large areas of National Forest Reserve land for planting eucalyptus, and reduced the maximum allowable area that could be rented from 2,000 rai to just 50 rai. The blow dealt by the Anan government's decision crippled the forestry policy and increased the pressure on the forestry bureaucracy. Foresters viewed the 1992 ruling as the main obstacle to the success of private tree plantations, and continued to lobby for politicians to have it reversed.[159] In January 2008 Deputy Director General of Forestry Upai Wayupat expressed the hope that the new government of Samak Sundaravej would revise this ruling and once again allow the private sector to lease degraded state forest land to set up plantations. "As soon as the new government comes into action and at the appropriate time, the RFD will present this proposal for plantation revision to the Cabinet."[160]

Industrial Strategies and Resistance

It's going to be camaldulensis *clones on good soil that would be better used for growing fruit trees and vegetables.*

Danish forester Jens Granhof on Kitti Damnern-
charnvanich's comeback[161]

With the 1985 forestry policy in place, Thailand attracted investments from powerful local and international firms. For pulp industrialists looking to plant eucalyptus in Thailand, the main challenge, simply put, was to get rid of farmers living in Forest Reserve land. These companies devised a variety of schemes to achieve this. The strategies adopted by four of them—Shell Thailand, the Thai-Japan Reforestation and Wood Industry Co. Ltd., the Finnish Jaakko Pöyry Oy, and Advance Agro (alias Suan Kitti)—provide an indication of the level of public resistance and the sheer difficulties of wresting land from people who have nowhere else to go.

The Suan Kitti eucalyptus plantations opted in the late 1980s for a combination of buying what was legally for sale, renting out small bits of government land, and paying non-tenured inhabitants to move away. The strategy chosen by the owner, Kitti Damnerncharnvanich, was an explicit attempt to avoid the mistakes of Shell Thailand. In Kitti's assessment, Shell had been greedy and hasty. "They were too impatient. They wanted to get the whole [plantation area] all at once in one single district!"[162]

Shell Thailand

By 1988 the Royal Dutch Shell Group had obtained huge areas of land in several countries for establishing pine and eucalyptus plantations: 100,000 hectares in Brazil, 53,000 hectares in Chile, 25,000 hectares in the Congo, 37,000 hectares in New Zealand, and 8,000 hectares in South Africa.[163] The company's original plan in Thailand was for a

125,000-rai (20,000-hectare) eucalyptus plantation in Chanthaburi's Khun Song Forest Reserve. However, when foresters surveyed the area in 1987 and discovered that more than half was intact forest, the project area was downsized to 10,700 hectares.

The site held the advantage of being close to the Laem Chabang deep-sea port in Rayong, and the three thousand people who lived in the area had no title at all. Shell was looking to invest 1.8 billion baht and hoped to sign a thirty-year contract with the government. If eucalyptus could be harvested after five years, this could mean up to six rotations. Initially, wood chips would be exported to Japan, Taiwan, and South Korea. In the longer term, a pulp and paper mill would be set up in Thailand, for which an additional 64,000 hectares of eucalyptus farms would be needed. Shell admitted openly that it would buy out villagers even though they had no legal tenure, and promised to hire a thousand people as workers.[164]

In answer to concerns about the ecological impact of the project, the multinational's local subsidiary, Shell Thailand, claimed that the plantations would improve the environment. Citing the work of Thai forest researcher Pitaya Petmak, Shell claimed that eucalyptus would increase rainfall, raise adjacent crop productivity, and be "friendly" to wildlife, especially bees.[165] While other companies crudely cited booming wood chip demand in Japan and huge potential profits, Shell made full use of the 1985 National Forestry Policy's environmental language, insisting that the firm's deeper purpose was to assist the Thai government with reforesting the country. It would also provide landless people with a secure source of income as plantation laborers living in company-owned "forest villages."[166] In fact, the Khun Song project was a non-native, single-species tree farm that would further expand Shell's already considerable cellulose investments worldwide. It would also be 100 percent foreign-owned and 100 percent for export. "Now that's good business!" remarked a Finnish embassy official in Bangkok.

The rules for renting out forest reserve land for large-scale planting in 1988 still stipulated that a company could obtain permission from the minister of agriculture for plots of up to 2,000 rai (320 hectares).[167]

Because Shell had opted to request the whole plantation area (10,700 hectares) at once, cabinet approval would be required before the project could proceed. The company was confident that what it had accomplished in Brazil, Chile, Congo, New Zealand, and South Africa could also be pushed through the system in Thailand. After all, Shell had broad international experience, a globally recognized trademark, and a managing director, Sarisdiguna Kittiyakara, who was none other than the brother of H.M. Queen Sirikit of Thailand.

At a time when the Nam Chon dam debate was in full swing and genuine environmental concern was on the rise, however, the Khun Song project became a highly controversial affair.[168] In 1987, following the incident at Non Lan in Si Sa Ket Province, orchard growers living in the Shell project area and Bangkok University students submitted a petition letter to the minister of agriculture. They expressed opposition to the project and voiced concern that large-scale eucalyptus cultivation could affect water flows from the catchment area as well as adjacent fruit orchards.[169] A few months later, dozens of rambutan and durian growers were arrested on charges of "illegal forest encroachment," and others complained that foresters were using "scare tactics" to get them off the land.[170] Sources in Chanthaburi said that local land speculators were paying poor villagers to clear away good forest in order to expand the permissible planting area for Shell, and eventually sell the land to the company. This suspicion was supported by Police Lieutenant Prathin Santiprapob who investigated the case. He claimed to be in possession of land satellite images that indicated unusually rapid clearing of forest in the Shell project area.[171]

As criticism grew in Thailand, and eventually in Britain and the Netherlands as well, Royal Shell Dutch hired the London-based International Institute for Environment and Development (IIED) to review the ecological and social impacts of the project. Environmental impact assessments (EIA) have never been required for industrial tree farms in Thailand. This study would not be an EIA but would provide evidence of the company's good intentions, Shell maintained.

IIED's preliminary report proposed measures to mitigate the damaging effects of monoculture in this rich fruit-growing region:

a version of *taungya* forestry, whereby farmers would be
   permitted to plant between the rows of young saplings
   during the first years in order to enlist their help in forest
   fire and weed protection and to avoid social conflicts;
plant different fast-growing species and inter-crops with
   farmers' cash crops as a way of avoiding disease and
   pest infestation that tend to attack monoculture
   stands;
leave "corridors" of native species around streams and on
   highly sloped hills; and
plan roads in such a way as to minimize soil erosion.[172]

To ensure success, the consultants also proposed that Shell should own or control well over half the plantation area. Acknowledging local people's unwillingness to be removed from their homes, they assured the Thai public that Shell would not engage in forced eviction.

Though the IIED study accepted the Shell plantation as a *fait accompli*, its conclusions clearly went farther than the company had expected. Shell had no intention of turning a straightforward, lucrative tree farm project into a time-consuming social contract with hundreds of villagers planting different cash crops in Shell land. The company's brief response to the report was that those measures that "did not affect the project's profitability" would be adopted.[173] One Shell employee hinted that if things became too complicated, the firm would move the project to Laos or Indonesia.

When Suan Kitti's plantation workers were arrested in January 1990, Shell was still waiting for cabinet approval. It never came.

Thai-Japan Reforestation Co.

From an investor's point of view, trying to set up commercial tree farms in National Forest Reserve land was fraught with political and financial risk. How could a firm get around the quagmire of the Thai forestry bureaucracy? It was a Thai forester who suggested to the Thai-Japan Reforestation and Wood Industry Co. Ltd. (TJR) that it could avoid these hassles altogether by dealing only with tenured farmers. Dr. Reungchai Pao-sujja of the government's Private Reforestation Promotion Office, the very same "godfather of *camaldulensis*,"[174] who had been the first to test this variety of eucalyptus in Thailand, advised TJR to "change direction." Stop seeking potential plantation sites in state land, he said. Focus instead on degraded tenured land. "Originally, [TJR] also wanted to plant in forest reserves. But they realized that permission was difficult to obtain and no one was succeeding... even Shell's project hasn't gone ahead yet," Reungchai commented shortly after the Suan Kitti arrests.[175]

In 1987 Japan imported 14 million cu m of chips for pulp production, of which about one third, or 3.2 million tons, came from Australia. TJR was formed, according to its managing director, Narong Supsuwan, as a direct response to environmental protests against industrial forestry practices in Australia, which created uncertainty about this vital raw material supply. As Narong put it: "How can we be sure that the Australian government won't cancel logging like they did in Thailand?"[176] The company was a local subsidiary of a consortium consisting of fifteen major Japanese pulp and paper companies, including Oji Paper, Jujo, and Mitsubishi Paper. Initially, TJR's projections were stunningly optimistic. It planned to plant an area of eucalyptus trees one hundred times larger than Shell's project: 2,000 sq km (more than 1.2 million rai). This would produce 2.6 million tons of chips, all to be exported to Japanese mills.[177] It would accomplish this by having two hundred thousand families in nine key eastern and northeastern provinces located within 300 km of the eastern seaboard ports plant eucalyptus in their own fields. Each family would plant about 5 rai of trees, producing 20–25 tons per rai after four or five years.[178] A contract would be drawn up between the company and planters grouped together in "forest cooperatives," obliging farmers to sell all the wood to TJR at a guaranteed price.[179] (Notably, there was no corresponding income guarantee for villagers in case of crop failure. In other words, Thai farmers would take the risk.)

In theory, farmers were to be encouraged to plant only on "degraded" farmland that would otherwise have gone unused. Reungchai insisted that farmers would not be allowed to plant eucalyptus on "good land," but conceded that ultimately they were free to choose what they wanted to do with their farms. In reality, villagers in Surin Province, where the first co-op was established, were coaxed to plant the fast-growing tree with promises of huge profits and negligible ecological side effects. A villager interviewed said that district agricultural officials "came to our village and explained to us that we should plant eucalyptus. They also said that a factory had been set up in the province which could assure us of a market."[180]

Farmers proved much less willing to plant the fast-growing Australian native than investors had hoped. Moreover, a co-financing arrangement, involving the Japanese aid agency Overseas Economic Cooperation Fund and the Thai Bank for Agriculture and Cooperatives fell through. As a result, only a small fraction of the projected 2,000 sq km were planted and the firms were reportedly exploring investment possibilities in Cambodia, Laos, and Vietnam. Though the TJR scheme never really took off, Oji Paper would a few years later buy its way into the Thai pulp plantation market through shares in Advance Agro, owned by Kitti Damnerncharnvanich.

Jaakko Pöyry Oy

Like Shell, Jaakko Pöyry Oy was in the late 1980s looking to expand the company's business activities into Southeast Asia. The Finnish firm's consulting division, at the time the world's largest forestry consulting firm, had a long track record of providing engineering, planning, and technical services to hundreds of logging, pulp mill, and plantation projects in Brazil, Indonesia, Australia, Portugal, Chile, Sri Lanka, Nepal, and Malaysia. But by 1985 the President of Jaakko Pöyry Consulting, Jouko Virta, who had been working on establishing a presence in Thailand, was becoming frustrated by the sluggishness of the Thai bureaucracy. A few months after the new forestry policy was announced, Virta had sent a Jaakko Pöyry representative to a Board of Investment seminar in Bangkok to try and generate interest in a plan for setting up tree plantations and pulp mills in Thailand. An article in *The Nation* entitled "Finnish firm studies Thai paper industry: suggests ways to cope with future demand," described the seminar.[181] There, the Pöyry representative proposed that Thailand would need 2,500 sq km of tree plantations to supply raw material to four or five new pulp mills that would produce 1 million tons of paper materials. The idea failed to arouse sufficient interest. Try as he might, Virta was unable to make inroads into the labyrinth of the Thai government system. "Business here is done on much more of a personal basis," he wisely concluded later in an interview.[182] Then Finnish forestry fortunes changed.

Virta was in Nepal in connection with a Finnish aid-funded forestry master plan being carried out by his company. Such "master plans" for the forestry sector were one of the company's specialties, which it produced in numerous tropical countries around the world with aid financing from the Finnish government. While in Kathmandu, Virta came across Nat Inthakan who was originally from Switzerland but had lived in Thailand for several decades. Nat had Thai nationality as well as a Thai name, and was well connected in Thai forestry circles, having worked for years at the sawmill in Sriracha.[183] Upon learning of Virta's frustration about getting a "foot in the door," Nat told the Finn with confidence that he could provide a personal escort to Thailand, and he guaranteed that meetings with all the necessary players could be arranged.

As he tells it, Nat Inthakan accompanied Virta to Bangkok later in 1986 and introduced him to three key bureaucrats: Dr. Snoh Unakul, then secretary general of the National Economic and Social Development Board; General Han Leenanonda, then minister of agriculture and cooperatives; and Phairoj Suwannakorn, then deputy director general (soon to become director general) of the Royal Forestry Department. In preparation for his trip, Virta said, he wrote up a terms of reference for a forestry master plan, which formed the basis of his discussions with the three government officials. "[The terms of reference indicated] how I conceived the Master Plan to be implemented in Thailand ... It has not changed significantly since then," Virta said.

What started out as a great business idea needed, in order to succeed, to be turned into an official development assistance project. Thus while the content may have remained the same, the packaging was improved. From a wily business plan, it came to be described as a Finnish *response* to a *request for assistance* by the Thai government. Conveniently, in 1988 Prime Minister Prem Tinsulanonda visited Finland where, as the project's final terms of reference put it, "it was agreed that such a project would be undertaken." Before Prem left Bangkok, Nat Inthakan recounts that he privately briefed the prime minister on the master plan concept. Nat was by then acting as local representative for the Finnish company.

A year later a new subsidiary, Jaakko Pöyry Thailand Co. Ltd., was formed with Nat Inthakan at its helm. But according to Virta, "this company is only a name. We need a name to operate in this country. There is a possibility that we would expand our consulting and engineering business here . . . one day we might establish a Thai group."

As for the actual Thai forestry master plan, it was Finland's first "aid" project in Thailand. While Denmark's long historical ties with Thailand help to explain Danish involvement in Thai forestry, Jaakko Pöyry's entry on the scene was more of a fluke. The master plan exercise was carried out by the Finnish forester and employee of Jaakko Pöyry, Rauno Laitalainen, who had just completed a master plan in Nepal and would, after Thailand, move on to assist Laos and then Bangladesh with more forestry planning. Finnish aid officials still to this day insist that the project was Finland's way of offering to Thailand its best know-how in a sector where Finns are world leaders. Whether Finnish bureaucrats are deluded or just naive, the project was a smart commercial move on the part of Jaakko Pöyry. (This transformation of a commercial scheme into a development aid project begs comparison with the same company's shenanigans in Ban Wat Chan). Though the forestry master plan was widely criticized by NGOs and community organizations for its focus on industrial forestry—and even a suggestion that logging should be started up again—the debate, which raged for several years, was largely shadow play for the company. It was far more important that the project provided a few years of open access to the opaque Thai forestry bureaucracy, during which ropes could be learned and alliances formed.

The subsidiary company Jaakko Pöyry Thailand never came to much; nor did the forestry master plan. But it did serve as a springboard for the Helsinki-based firm that went on to carry out several commercial contracts for Thai pulp and paper interests, including Siam Pulp and Paper, Phoenix, and Advance Agro,[184] as well as numerous others in the energy sector in Thailand and the region.

Kitti's comeback

The 1990 arrests of Kitti Damnerncharnvanich's plantation workers marked a turning point in Thai forestry. It prompted a drastic reduction in the allowable size of plots that could be leased for commercial tree planting, making it more difficult than it had been to rent out state land for establishing industrial eucalyptus farms. Kitti had legally rented some 110 sq km of state forest land and controlled another 160 sq km. These plantations provided raw material for two pulp mills in Chachoengsao operated by Advance Agro, Kitti's giant pulp and paper concern that was co-owned by Finnish Stora Enso and Japan's Oji Paper.[185] But as more people experienced the impacts of eucalyptus on groundwater and surrounding farmland, local protests against tree farms continued.

Wiboon Khemchalerm, the famous organic farmer in Chachoengsao Province, warned that many villagers were being forced to accept money from the company and to sell their rights to the land. "Once the money is gone, and there's no more land to work on, where will these people go?" he asked.[186] Lum Jumchai, a sixty-year-old single mother of ten children, fought a six-year legal battle with the Soon Hua Seng Group, which owns Advance Agro, to reclaim 4.5 hectares of land in Laem Khao Chan village, Chachoengsao Province. When she refused to part with her land, the company filed a lawsuit for encroachment on private property and provincial police repeatedly threatened to dismantle her house. Assisted by lawyers from the Bangkok-based human rights group, Union for Civil Liberty, Lum Jumchai finally obtained a court ruling asserting rights to her land.

"I plant cassava now. After the court order, the company does not harass me anymore," she said. But she faces problems with her cassava crop because the surrounding eucalyptus plantation is drying the underground water sources and hardening the soil. And in spite of Thai foresters' steadfast support for eucalyptus, the industry had to devise new strategies to secure access to more land.

Seen in this perspective, Kitti's plan to bring in massive Chinese capital and political clout to establish his Sino-Thai eucalyptus plantation reflects as much the undiminished ambitions of its owner as a measure of what was needed to overcome the resistance of Thai farmers.

The US$1 billion plan to establish a 1,200 sq km eucalyptus plantation in Thailand for export to China was first announced in 1997. Ownership would be divided between the Chinese government (51 percent) and Kitti's pulp and paper firm, Advance Agro (49 percent). It was billed as China's single largest overseas investment.[187] According to the chair of Advance Agro, Virabongsa Ramangkura, the firm was chosen among competitors in the region because it said it could build the plant in less than two years, compared to the three or four years required by its competitors. For more than three years, there were numerous comings and goings at the highest levels of government between Thailand and China, including a visit by Chinese President Jiang Zemin to Thailand and a 180-person Chinese delegation in 1999. An official agreement was signed a few months later.

Like all the others though, the problem of this scheme was finding the land to plant the trees. At the time, director general of the Royal Forestry Department Plodprasob Suraswadi was hopeful (because of Kitti's "good contacts" with the Chinese) that the Sino-Thai project would succeed. He even viewed it as a way of helping "to create pressure to change Thai forest laws" such that much larger areas of state land could be rented out. In a revealing interview, Plodprasob noted that Kitti wanted to use 320 sq km (200,000 rai) of his own land and 800 sq km (500,000 rai) of land planted through contract farming arrangements with villagers, both those with land tenure and those without.[188] Plodprasob was willing to overlook the fact that half of Kitti's "own" land was, in fact, plantations in state land with no legal basis. Would this not imply legalizing illegal land holdings in state forest land? "I don't want to call it legalizing," Plodprasob replied. "But they already own it illegally."

The director general also looked favorably on the fact that the Forestry Department would not have to engage in negotiations with farmers directly; the company would "deal with the people themselves." Clearly viewing the department in a position of advantage now, Plodprasob had even placed conditions on Advance Agro. "I told Kitti straight out: You have to plant me 100,000 rai of permanent forest... That way we get the forest back. This is non-negotiable," he said. Kitti would, in addition, have to establish an environmental fund so that the director general could use the money for "other projects that might come up."

It's not clear that the fund or the "permanent" forest ever materialized. But Advance Agro's eucalyptus plantations continue to dominate the eastern Thai landscape and, within the Thai forestry leadership, there remains the hope that state foresters will regain a role in overseeing these vast tree farms.[189]

The End of the Road

> *We've brought our thinking to the tropics, but we are very influenced by the relative simplicity of the boreal and temperate ecosystems . . . That's why it is so tragic to use this technique in the tropics. From the ecological point of view, adopting a . . . Nordic [forest management] system is just plain dumb.*
>
> Olle Zackrisson, Swedish forest biologist[190]

Backed by the juggernaut of colonial expansion, foresters introduced a similar forest management "formula" in colony after colony. In the development era, principles of scientific forestry have been stretched to their logical conclusion with industrial planting of non-native fast-growing species in vast tracts for a single purpose: wood production. The hierarchical, militaristic culture of foresters has tended to suppress internal dissent, thus helping to maintain the uniformity of the approach.

This may also help to explain why a paradigm shift in Thai forestry has been so slow in coming.

While Thai farmers' protests against industrial tree farms continue, Thai foresters have for the most part not studied—or even questioned—the long-term environmental impacts of large-scale eucalyptus plantations in any serious way. Through their omission, they condone a colossal ecological experiment whose negative consequences could in the end far outweigh the profits to those private firms that have managed to access land for commercial tree farms.

But if Thai foresters cared to look, they would find that generations of experience with monoculture tree planting in the West—from Germany, the very cradle of global forestry, to Scandinavia to the United States—provide a glimpse of where it could be heading.

While "the short-term success of [the German] forest conversion was and is extraordinary,"[191] inherent flaws in the approach to forest management have become increasingly apparent with time. Man-made forests tended to have weakened resilience to stresses like wind and fire, outbreaks of pests tended to be more frequent, and 20–30 percent production losses in later rotations have been observed.[192] As Munich University Forestry Professor Richard Plochmann wrote in 1968: "It took about one century for [the negative biological consequences of intensive forestry] to show up . . . Many of the pure stands grew excellently in the first generation but already showed an amazing retrogression in the second generation."[193] The term used by German foresters to describe this widespread phenomenon is *waldsterben*—forest death.[194]

Today recognition of these fundamental shortcomings has forced a complete turnaround in German forestry. German forest owner Hermann Hatzfeldt gives the "final demise" of conventional forestry an exact date: February 28, 1990, the day an intense wind storm, "*Wiebke*," wrought havoc all over Germany. "The damage was predominantly in older single-species, even-age stands," Hatzfeldt writes. "As man-made trees-in-a-row forests were crushed in one instant, so were the hopes, ambitions, and high pretensions of foresters and forest scientists . . . After *Wiebke*, traditional forestry came to an end in Germany."[195]

German forest policy now embraces, in principle, the concepts of ecological forestry, and forestry education is undergoing a fundamental reorientation. Freiburg University in Germany was the first place in the world where scientific forestry began to be taught in the late eighteenth century. Yet from 1995 a "radically revised" forestry curriculum was introduced at Freiburg that does not prioritize timber production above all other forest values. According to one of the architects of this historic curriculum reform, forestry professor Siegfried Lewark, forestry research and teaching now focus not exclusively on timber production, but also on the "natural laws of forest ecosystems," and the significance of forests for climate, soils, air, water quality, and recreation. Forestry education puts emphasis on society's and forest owners' objectives for forests, and on the political relations among society, forests, and forestry.[196] The new approach, Lewark says, encourages students to be "less passive absorbers of information and to be more active, creative, good at working in groups, inquisitive, and to have initiative."[197]

Asked about the application of "conventional forestry" in the tropics, foresters like Lewark warn that impacts are likely to be even more serious because of the higher levels of biological diversity. "Temperate forests are possible to regenerate. On the level of tree species, there are only twenty or thirty in Central Europe. We have a very different ecological situation here," he says.

In the United States, similar changes are also slowly taking place. Utah State University forestry professor Jim Kennedy describes the early stages of American forestry where "ecosystems were often conceived and managed like Henry Ford's early automobile production lines . . . The management approach was usually reductionist . . . Complexity and diversity were commonly viewed as the enemy! We foresters were often taught to deductively simplify, compartmentalize, and dominate nature (both our own emotional, intuitive, playful human-nature, and the natural systems around us)."[198] Moreover, discipline within the forestry hierarchy, he says, was maintained through a "one-way patriarchal flow of control from male line managers." As the ecological perils of this "machine-model view" of forests become more evident, Kennedy says, American foresters have to challenge their own traditional assumptions to accommodate the "diversity, complexity, and interdependence of ecological and human systems."

Throughout Northern Europe, German-inspired forestry practices adopted over a century ago are also changing. In Denmark, as with Germany, windstorms during the 1990s were decisive in altering the direction of Danish forestry. After a storm in 1999 that laid waste to vast areas of forest, especially single-species plantations, the director of the Danish Center for Forest Landscape Planning, Niels Elers Koch, recommended that the Danish government adopt a forestry that creates as much variation as possible and uses local species. "If you have one clone on a huge piece of land it is a very risky affair. I believe we need to look at different values in each different area. No forest should have one single goal of wood production." Reflecting on the approach of classical forestry, he asks: "Are there really areas where we don't want to protect the groundwater and the flora and fauna? In the long term this is not very wise." Today Danish farmers receive government subsidies not to plant fast-growing conifers, but slow-growing deciduous trees like oak and beech.

Over the past decade, forestry in Sweden has undergone a similar transformation. The National Forestry Board, once charged with the sole task of ensuring maximum wood production, now focuses primarily on environmental matters related to forestry.

Forest biologist Olle Zackrisson has spent decades struggling against what he viewed to be the devastating ecological effects of Swedish forestry practices. He acknowledges that Sweden and Finland are world leaders in timber and pulp production, and that the forestry industry is largely responsible for these countries' rapid economic development. Indeed, it is this success that has caused so many tropical countries to look to the lucrative Scandinavian model for inspiration. But this success came at a high ecological cost. Since the Second World War, a number of measures were introduced to accelerate the process of replacing forest with nursery-raised seedlings in clear-cut plots. These included the use of chemical fertilizers, spraying of herbicides to destroy unwanted plants and tree species that competed for soil nutrients, and planting with foreign species and provenances such as the North American lodge-pole pine. Rules and regulations also forced forest owners to clear-cut and replace "low productivity forests" with fast growing tree plantations. A system of fines and grants penalized "passive" forest owners and encouraged an increasingly intensive style of forestry. Economic development was achieved through radical manipulation of the forest ecosystem, and its replacement with plantations of one or two tree species.

But dozens of animals including lynx, wolf, black stork, and white-backed woodpecker as well as hundreds of species of moss and fungus have become endangered or extinct in the region. By the 1980s, the Swedish public became aware of the ecological implications of this combination of methods and many forestry practices were eventually banned. Herbicides are no longer used, and problems with disease in lodge-pole pine have caused some to question the wisdom of such large plantations of this non-native species.

When he travels to the tropics, Zackrisson is constantly amazed at how fellow scientists and foresters view Scandinavian forestry. He often despairs to hear his country referred to as an example of "successful" forest management. "I feel so sad that we are exporting this silviculture to other countries, and people seem to believe it's the most modern thing."[199]

"As a young biologist, I started meeting [Swedish] foresters. They seemed to look on the natural forests as somehow decadent, rotten, negative ... even degenerate places." The task of foresters "was not to mimic nature, but to put order into nature, to make it less messy, more organized, more simple." As a result of the fierce public debate in Sweden, Zackrisson speaks of "the whole paradigm falling to pieces," and urges tropical foresters to beware. "[They] should realize that we all have to change now. They should change before they are standing at the end of the road like Sweden is today, where we have only 2 percent [of natural forest] left."

PART 4

THE MAKING OF THAI WILDERNESS

A study of conservation imperialism is long overdue.
Ramachandra Guha[1]

Just as scientific forestry was exported throughout the tropical world, the nineteenth-century American national park ideal—where "un-peopled" wilderness is cleared of local people and opened up for tourism—has had an equally extensive global influence. Both were adopted uncritically by the Thai state forestry bureaucracy during the twentieth century, with virtually no attempt to question their applicability or to adapt the ideas to local biological and cultural realities. Thai national parks and wildlife sanctuaries were created based on the core belief that within their boundaries there should be no human disturbance. What this has meant, in fact, has been excluding local communities while at the same time massively promoting tourism.

When the divisions for National Parks and Wildlife Conservation were set up within the Thai forestry department in the 1960s, they were like the bastard offspring of the forestry establishment; low priority units that generated no income and enjoyed little funding, prestige, or political support. As tree planting was to logging for much of the last century, conservation was initially viewed by mainstream forestry as little more than a cost. The cancellation of commercial timber concessions in 1989 changed all this. It stripped Thai forestry of its original, central purpose—to log timber—and the protection of "what was left" took on a new urgency. Indeed, as political obstacles continued to hamper the advance of commercial forestry, protecting natural forests gained importance and status within the Royal Forestry Department. The supremacy of the conservation agenda precipitated a transformation of forestry—a reversal, really—from an activity primarily preoccupied with cutting down trees to one focused overwhelmingly on protecting them. By the mid-1990s, the RFD's Forest Conservation Program had the largest budget of the department's six programs. The establishment in 2002 of a new, well-funded National Parks, Wildlife and Plant Conservation Department, which usurped much of the administrative and geographic territory of the old RFD, was a final, natural consequence of this evolution.

With the redefinition of forestry's mandate came an enormous expansion of the physical territory covered by conservation areas. The new boundaries have enclosed and overlapped with the settlements and forests of thousands of communities, many of them ethnic minorities, whose presence has become an intractable dilemma of Thai forest politics.

Theft and Utopia: The American Model

> *I am rocked in the ancient land . . . In the beginning of the [twentieth] century the Indian smoke still mingled with ours. The frontier of the whites was violent, already injured by vast seizures and massacres. The winter nightmares of fear poisoned the plains nights with psychic airs of theft and utopia . . . The severity of the seasons and the strangeness of a new land, with those whose land had been seized looking in our windows, created a tension of guilt and a tightening of sin.*
>
> Meridel LeSueur, *The Ancient People and the Newly Come*[2]

Uninhabited wilderness is not a physical place that was discovered by the Newly Come. The idea of preserving a vast wilderness in its pristine state—in the form of a national park—is a utopian vision projected onto conquered peoples and their landscape by the victors.

American national parks are predicated on the massacres, forced resettlement, legal battles, and racism that accompanied and facilitated the dispossession of native peoples. This is a bitter pill for many of us, given the pervasiveness and global appeal of the wilderness myth. Acknowledging the fact of dispossession reveals American wilderness to be *emptied*, rather than *empty* land. It forces us to rethink the moral, legal, and scientific justification for normative conservation dogma adopted in Thailand and many other tropical countries. This dogma holds that "in national parks there should be no human activity whatsoever."[3] At the very least, reflection on the origins of the wilderness concept is absolutely crucial for Thailand's contemporary "people and parks" debate.

In the 1860s, the United States had no thousand-year-old churches or classical monuments in its capitals, but it did have nature on a scale unmatched in Europe. "Wilderness," writes Roderick Nash, "had no counterpart in the Old World."[4] For an emerging nation in the midst of civil war, the idea of a wilderness, "where nature reigns in her virgin beauty,"[5] served as a powerful symbol of national identity and unity. It was a quintessentially American invention, which helped Americans define their new culture in a way that compensated for a short history, weak traditions, and minor scholarly and artistic achievements. Nature writing became a literary genre and the fame of wilderness spread through the poems, books, and newspaper articles of East Coast, urban gentlemen who visited the wild, untamed West and then went home, for the most part, to their desks in the East to sing its praises.

The classic study, *Wilderness and the American Mind* provides a fascinating account of the history and evolving meanings of nature in America. Yet scholars of American wilderness, Nash foremost among them, have tended to be silent about its one, obvious, central fact: that wilderness was created—both in the public imagination and in the landscape—by forcibly removing the peoples who lived there.[6]

Strangely, in the early part of the nineteenth century, conceptions of American wilderness did include Indians. The painter George Catlin, who is widely credited as being the first to articulate the national park idea, wanted to include forests, buffaloes, as well as Indians in "a *magnificent park*." "What a beautiful and thrilling specimen for America to preserve and hold up to the view of her refined citizens and the world, in future ages! A *nation's park*, containing man and beast, in all the wild[ness] and freshness of their nature's beauty!"⁷ Catlin's Indians were, however, not peoples with complex histories and politics, as caught up by the dramatic changes sweeping the continent as were the immigrants from Europe. Rather, in keeping with his romantic ideals, Indians were tragic "children of nature," noble, innocent, and "fresh," as wild as the forests they inhabited and inextricably linked to a threatened landscape. For Catlin, these classic natives had, like the trees and the mountains, always been there.

Some groups that Catlin observed in the Mississippi valley were indeed anciently indigenous. But on the whole, the situation for native peoples there and in the rest of the continent was highly fluid. Eastern tribes were being forced westward and groups from the Great Lakes region were pushed south. Some adopted equestrianism such as the Sioux, Comanche, and Shoshone had earlier done, and left their villages for a nomadic life of hunting on the plains. Peoples were squeezed onto ever-smaller parcels of land, and countless died from starvation, wars, and disease. In any case, the original American idea of parks with people in them is largely forgotten. By the 1850s, the romance of the exotic savage inseparable from his natural forest home was eclipsed by another, equally idealized notion: uninhabited wilderness.

With legislation signed by President Abraham Lincoln in 1864, a 16 sq km patch of forest in Yosemite Valley became the model for national parks throughout the United States. Indeed, the basic formula—fencing off an area and removing its inhabitants in order to protect nature and provide a place of recreation for urban tourists—would be copied all over the planet. In spite of innumerable cultural, ecological, and political differences between two countries on opposite sides of the world, and not least the giant gap of a century of time, Thai protected areas were in the 1960s based quite explicitly on this American model. "Actually, all over the world," observes Thai forestry professor Surachet Chettamart, "on every continent, Yosemite and Yellowstone [national parks] have been used as models for conservation, including in Thailand."⁸

Today Thailand's protected areas system, which covers more than a fifth of the land area, is not functioning. National parks are being overrun by too much tourism and are threatened by all manner of private- and state-funded projects. And while Thai conservation laws state that no one is permitted to live inside national parks and wildlife sanctuaries, at least 2 (or more) million people, mostly ethnic minorities, in fact, inhabit these parks. These communities live with the constant threat of eviction hanging over their heads, and foresters are, strictly speaking, obliged to remove them. It is no wonder that Thailand's protected areas are plagued by conflicts with local people.

Foresters like Surachet, who studied conservation in the United States, realize these shortcomings. He has pointed out that "the problem of just copying US law without looking at the situation in Thailand is that it ignores social and cultural realities. Politically speaking, we are still sticking stubbornly to the US model so there's always a gap between law and reality."

His assumption, a common one in Thai forestry circles, is that while the world's first two national parks are prototypes for preserving a *pristine* nature, Thai national parks have not succeeded because farmers—driven by poverty, ignorance, and sheer numbers—invade park boundaries, and state authorities. Due to lack of funds, corruption, and general ineptness, they are unable to protect park lands. While all this may be partly true, Surachet's assumption is fundamentally flawed. The shortcomings of the Thai national parks system are not a result of a failure to apply the principles of conservation. The Thai forestry bureaucracy has arguably, in its antagonism towards forest-dependent people and its relentless efforts through the years to have them evicted, been relatively faithful to the spirit of the American conservation model. Rather, the intense social strife and at times heavy-handed measures used against local people are an *inherent* part of the American national parks model. The sad tales of the founding of Yosemite and Yellowstone and the violent clashes with their inhabitants are eerily prescient of the conflicts between local people and forest conservation authorities in contemporary Thailand.

Yosemite

John Muir was one of the American conservation movement's most influential thinkers. He first stepped foot in the Yosemite Land Grant four years after it was established and became the area's most vociferous advocate.[9] A mountaineer and prolific nature writer, he made the Californian valley famous through his prose and lobbied for its expansion and protection from loggers, miners, and settlers. Muir was as disgusted by the rampant commercialism of his age as he was captivated by the awesome beauty of Yosemite's sheer cliffs, rivers, and forests of millennial trees. Anyone who has stood beneath those same cliffs will recognize Muir's exhilaration. "God is in it,"[10] he stated simply. "I am lost—absorbed . . . with the divine and unfathomable loveliness and grandeur of Nature."[11] Recounting his discovery of a group of small white orchids in the woods one day, which he would later describe as a moment of epiphany, he wrote: "I felt as if I were in the presence of superior beings who loved me and beckoned me to come. I sat down beside them and wept for joy."[12]

His opinions of the people who had until very recently called this landscape home stand in sharp contrast. Muir was clearly aware that Indians had lived in the park. But they were incompatible with his way of seeing nature. They were, in his words, "mostly ugly, and some of them altogether hideous." Writing in 1894, he observed that Indians had "no right place in the landscape."[13] As one offended upon waking by the rude intrusion of reality into a remembered dream, Muir could not feel the "solemn calm" of wilderness when Indians were around.

From the bulk of his writings one would never suspect a native presence in Yosemite. Yet the ground that produced Muir's transcendental revelations had been home to Ahwahneechee and Miwok ("Digger") Indians. The latter constituted one of the largest tribes north of Mexico and possessed elaborate land use and tenure systems.[14] Muir's "virgin land illusion" seemed true because only a few decades earlier most of the original inhabitants of the valley had been killed or driven out.[15] Indeed, the hundred years that preceded the creation of Yosemite witnessed an 80 percent decline in the native population of California. The discovery of gold in 1849 then led to the destruction of the forests and streams upon which people depended, and contact with miners resulted in severe epidemics that weakened the tribes and sparked off violent confrontations. It has even been suggested that Yosemite's native population initially increased after its 1864 designation as a conservation area as Indians sought refuge from surrounding regions.

The very name of the valley reflects its bloody history. According to one theory, the leaders of the native Ahwahneechee were reportedly so fierce in battle that they claimed kinship with the grizzly bear, in their language, the *uzumaiti*.[16] They waged a quixotic war against the white gold miners who encroached on their game fields and acorn orchards. But the Mariposa Battalion led by Major James Savage in 1851 punished those "who had not proven totally submissive when miners had expropriated their land."[17] The soldiers pursued the Ahwahneechee to their hideaway and thus "discovered" the valley. The Indians were removed but the grizzly-bear name, written as Yo Semite, and eventually Yosemite, remained.[18] A second theory about the origin of the name holds that another group, the Miwok, who lived west of the valley, referred to the Ahwahneechee as *johemite*, which can be translated as "some of them are killers." The armed men who entered the valley in 1851 may have learned of the Ahwahneechee village location from the Miwok, and so "the present name of the valley dates back to this invasion."[19]

The forced evictions of 1851–52, which settlers referred to as the Mariposa Indian War, were accomplished through lynching and slaughter, burning of villages, and starving or freezing people into submission. In 1890, the year Yosemite was expanded and officially designated as a national park, the Ahwahneechee submitted a petition to the US Congress demanding reparations to compensate for the victimization of their families a half century before. The petition was ignored.[20]

Yosemite park authorities tolerated a limited Indian presence for many decades because labor was needed to service the booming tourist industry. Native men chopped wood and worked as bellboys, guides, and drivers, while women were employed as domestics in private homes and as hotel maids and laundresses.[21] They also earned income by selling baskets, strawberries, and fish to park visitors in an arrangement that might today be described as "eco-tourism." Nevertheless, to protect tourists against the perceived threat from Indians, the US Army ran the park for its first fifty years. By the late 1920s, with a preservationist ethic taking root in the public imagination based on "unimpaired" areas, a master plan was prepared for Yosemite that paved the way for the final removal of native people from the park. The last Indian dwellings in Yosemite were burned down in 1969.[22]

Yellowstone

Yellowstone's founding story provides another prototypic example of removing native populations in order to create *uninhabited* wilderness. Throughout the park, there is archaeological evidence of the widespread use of fire and of ceramic pots, arrowheads, and ancient ceremonial sites that suggest a human presence dating back as far as seventy-five hundred years. Inhabitants of the area mined, worked, and traded obsidian (a type of stone used for making tools and weapons), they hunted and gathered, and they established a system of trails upon which the park's modern highways are based.[23] The regular, controlled use of fire created and maintained plant and animal habitats on which people depended, and helped to prevent the massive burns that now annually plague the western forests, where a policy of fire exclusion has been adopted.[24]

By the 1800s, the area was home to both equestrian and non-equestrian Shoshone, Bannock, and Mountain Crow peoples, while it was used seasonally by several other groups, including Blackfeet, Nez Perce, Salish, Kalispel, and Coeur d'Alene. The 1851 treaty between the US government and native leaders recognized territorial rights of the Shoshone, Blackfeet, and Crow nations that covered most of the area that eventually became the park. The willingness at the time to grant such extensive rights reflected the relative political strength of native peoples. The government was also concerned about placating the tribes in order to guarantee the safe passage of hundreds of thousands of settlers heading west toward Californian gold. In the ensuing years, as European Americans consolidated control over the continent, the tribes were forced to cede much of this land. But they did retain the right to hunt on the so-called "unoccupied lands of the United States." That right would also finally be extinguished by a US Supreme Court ruling in 1896.[25]

Though trappers had traversed the Yellowstone region for many years, one of the first official parties of white men to explore the area, the so-called Washburn expedition, took place in 1870. The party was funded by the Northern Pacific Railway Company, which had interests in creating a tourist destination for train passengers from the east.[26] It included prominent Montana citizens who were able to afford a heavily armed military escort "to protect against possible unpleasantness with the Crow Indians."[27] The party benefited from the use of Indian trails through dense forest, came across numerous native encampments, and even bumped into a large group of Crow hunters and their families. Tellingly, they also witnessed a "surround burn," a small, controlled fire set by Crow hunters to encircle game with a ring of fire. Yet these brave explorers still managed to conclude after their three-week journey of discovery that Yellowstone was a primeval wilderness "never trodden by human footsteps,"[28] and they lobbied for the preservation of this "natural wonderland." It took just two years; Yellowstone became a national park in 1872.

Violent conflict over this spectacularly beautiful area continued for longer than in Yosemite. In 1877 some two thousand troops pursued 750 Nez Perce men, women, and children for ten weeks, two of them spent inside the park where they reportedly accosted a few parties of tourists. Their desperate retreat finally ended south of the Canadian border. In 1878 the army was again called in to quell an uprising of several hundred Bannock. Consequently, when the first park headquarters was established the following year it was a "heavily fortified blockhouse" designed to handle the "Indian troubles."[29] Yellowstone's first superintendent, Philetus Norris, lobbied for the expulsion of local people "averting in future all danger of conflict between these tribes and tourists."[30] He was also concerned that Indians in the park threatened to jeopardize pending leases for the construction of hotels and other tourist facilities. In spite of his obvious awareness of the native presence, which continued for several more decades, Norris contributed actively to the myth that Yellowstone had always been un-peopled. He did not say: we won the war, now you have to leave. Instead, he argued to the government that Indians had, in fact, never used the park because they were afraid of the hot, steaming geysers: "The isolation of the park... and the superstitious awe of the roaring cataracts, sulfur pools, and spouting geysers over the surrounding pagan Indians, [caused them to] seldom visit [Yellowstone]."[31]

Notwithstanding the false geyser taboo that helped justify the park's existence, native people did frequent the geysers—and continue to do so—particularly for religious purposes.

In the early years of the park, tourists carried guns for protection and for hunting. Road building had to be accompanied by well-armed men, under constant threat from Indian raiders. Carrying firearms was not forbidden until two decades after Yellowstone was designated.[32] In 1886, as with Yosemite, the US Army took over complete control of the park. Captain Moses A. Harris, a frontier cavalryman, Civil War hero, and veteran of several campaigns against Indians, became superintendent. His approach toward native people at the close of the nineteenth century foreshadowed the attitude of Thai forest conservation authorities at the dawn of the twenty-first century. He maintained that "the Government would not tolerate hunting, fishing, destruction of timber... within [Yellowstone and would] adopt effective measures to keep the Indians... away from the Park." He was prepared to have the aid of the military to remove them and to administer "proper measures ... for their punishment."[33] Nature conservation remained in the hands of the army to protect vacationers from attack by "hostile savages."[34] After Harris, twenty-one more soldiers would have responsibility for the park before it finally came under civilian control.

Far from being anomalies in their treatment of native people, the United States' first and internationally best-known parks, Yosemite and Yellowstone, provided models for policies of Indian removal that were used throughout the country. Notwithstanding the enduring myth of unspoiled wilderness, all the largest and most exquisite American national parks—the so-called "Crown Jewels" of the US parks system—had long histories of human presence, were most definitely inhabited, and had to be *emptied* of people. Yosemite, Yellowstone, Mesa Verde, Mount Rainier, Olympic, Grand Canyon, and Glacier national parks were all founded upon the same fundamental contradiction: They were created to preserve uninhabited wilderness, yet necessitated the forced eviction of prior occupants, and ignored their territorial claims and sacred sites.

The general silence surrounding these blatant injustices can be partially explained by the elitism and even racism within the wilderness preservation lobby. The American conservation movement began as the preoccupation of the privileged class, who in the early twentieth century were by definition people of European Protestant background. But long after the upper echelons of American society became more diversified, members of conservation organizations "retained [their] ethnic homogeneity." US conservation historian Stephen Fox observes that well into the 1960s, the movement continued to be dominated by an "old WASP gentry."[35]

Decades later, relations between American national parks and Indian reservations remain troubled and complex. Some parks border land owned by one or several different tribes, some parks surround reservations, while there are also reservations that encircle parks.[36] All of the Crown Jewels (and several other parks besides) are contested by native people, who argue that the land was taken from them illegally and continue to this day to demand it back.[37]

George Ruhle and the Story of Glacier

> [This] is what I now give you... We will sell you the mountain lands from Birch Creek to the boundary, reserving the timber and grazing land... We don't want our Great Father to ask for anything more... If you come for any more land, we will have to send you away.
>
> Chief White Calf on the forced sale of Blackfeet lands, 1895[38]

Yosemite and Yellowstone have a symbolic importance for the situation of national parks in many tropical countries whose conservation systems are based on the American model. But Glacier, the eleventh American national park founded in 1910, has special relevance for the Thai forest story. It was here that George Ruhle, who helped draft Thailand's forest conservation laws, began his fifty-year career with the US National Parks Service. He was the first chief naturalist to be appointed in the United States, holding this position at Glacier in the turbulent thirteen-year period from 1929–41 when park officials were in constant conflict with native people.

Ruhle arrived in Thailand in 1959 and, after a brief tour of potential national park sites, came to the conclusion that "unassimilated non-Thai" hill people were primarily responsible for deforestation, and were, in fact, causing more harm than loggers. He did not cite any evidence for this claim, and he did not mention any other causes of deforestation. He merely stated this as fact after a few months in the country. Ruhle died in 1994 and I was not able to interview him. It seems reasonable to assume that his perspective on Thailand's ethnic minorities was influenced by the prevailing prejudices of the Thai government officials he met while he was traveling in the kingdom. It is also likely that he was carrying prejudices of his own. One cannot help notice the overwhelming similarities between Ruhle's assumptions about Thailand and the approach of Glacier authorities towards Blackfeet native people. Though Ruhle was not personally antagonistic towards Indian people and even encouraged a deeper understanding of their culture by visitors to the park, it was a matter beyond dispute to him that the Blackfeet claims to Glacier had no legal basis and that their hunting and gathering inside park boundaries constituted a threat to the park.

A brief look at Glacier may help to shed light on the thinking of the American conservationist who has been called a "Father of the national parks of Thailand."[39]

Prior occupants

Whereas open conflicts with prior occupants of Yosemite and Yellowstone parks had by the late 1920s largely subsided, Glacier was, in 1929 when Ruhle arrived there, in a "near state of war."[40] Blackfeet people who had been resettled onto a reservation to the east still claimed and exercised hunting and use rights in the eastern half of the park. Park managers accused the Indians of "wanton slaughter" of wildlife and attempted to ban them from using the park. Ruhle was never directly responsible for handling Glacier's "Indian troubles." He was in charge of visitor education and guiding, naming sites in the park, and documenting its natural features. Ruhle called Glacier "the most sublime wilderness in America." (According to park authorities, his guidebook to Glacier, first published in 1949, is still in use.) But Ruhle clearly shared Glacier management's attitude that native people were a threat to the park.

The Blackfeet Confederacy consisted of three linguistically related tribes—Pikuni, Kainah, and Siksika—that had migrated to the area in the early 1700s, driving the Kootenai and Shoshone further west over the Rocky Mountains. For more than a century, they controlled much of the plains, feared by the British, Americans, and other Indian tribes. Their current claims to land inside Glacier National Park date back to 1895, when the Indians were forced to sell the eastern mountainous part of the park but maintained the right to use these lands. The 1895 agreement with the US government covering some 2,000 sq km stated that Blackfeet people retained "the right to go upon any portion of the lands . . . to cut and remove timber for . . . personal uses . . . [and] to hunt upon said lands and to fish in the streams thereof, so long as . . . they remain public lands of the United States."[41]

A decade of lobbying by a coalition of wilderness preservationists, politicians, and railroad magnates resulted in the area becoming a national park in 1910. The legislation for the new park made no mention of Blackfeet use rights. Rather, it stipulated that since the park was like a giant laboratory of wildlife, glaciers, forests, and rare flora, everything must remain undisturbed in a "state of nature," in the interests of science, game protection, and tourism. According to a leading proponent of the park, George Bird Grinnell, Glacier was "absolutely virgin ground."[42]

The Blackfeet disregarded the new law, and continued to use the eastern mountains in the park for hunting and fishing, for gathering food and medicinal plants, and for religious purposes. These practices were, strictly speaking, illegal and resulted in endless conflicts with park authorities. On the other hand, the participation of Blackfeet in promoting tourism was actively encouraged. The Great Northern Railway Company that built the massive Glacier Park Lodge in the park arranged for native dancers to greet visitors from the east as they stepped off the train. Tourists also had the opportunity to photograph and to be photographed with real live Indians and to buy mini-teepees set up on the manicured hotel lawns. Blackfeet boys, dressed in full tribal regalia, enhanced the outdoor experience by carrying golf bags as tourists teed off. Glacier was like an early twentieth-century Disneyland where visitors were spoon-fed with pseudo-native kitsch.

Glaring double standards pervaded other aspects of park management. In spite of their best efforts to control Blackfeet use of the park, rangers often encountered native hunters inside park boundaries, and reported frequently on their "ruthless slaughter" of elk and other game animals. Ruhle charged that deer and elk had disappeared from the Indian reservation adjacent to the park as a result of over-hunting. "Where are they now? They're all gone," he reflected in an interview many years later.[43] Describing an encounter with the Blackfeet, he said to them: "But you can go up there in the park and see them, and when an elk comes over this boundary, he's yours. You can take him, we can't say anything about it. If there was not elk up there [in the park], there wouldn't be any coming down here [to the reservation]...When the day comes that you can come to me and say, 'We understand; it's better now the way it is, and we want to keep it that way,' then you and I won't have any differences."

The claim that the disappearance of game was caused by Blackfeet hunting must, however, be considered within the larger picture of the authorities' overall approach to wildlife, which appeared to place tourist enjoyment above all else. Until as late as 1945, four years after Ruhle left the park, Glacier's policy on wildlife was essentially to protect *desirable* animals. These—in a bizarre parallel with the modern forester's focus on commercial tree species—included elk, deer, bighorn sheep, and mountain goat, at the expense of predator species like wolf, bear, and wild cats, which were deemed less attractive and more dangerous to tourists. This was accomplished in two ways. Winter feeding of ungulates and mountain goats helped to increase the populations of these preferred species and thereby the likelihood that tourists would "spot" them. At the same time, park management actually killed off animals—coyotes, wolves, mountain lions, and even eagles—by poisoning or shooting them.[44]

A description by George Ruhle of a particularly enjoyable trip he led into the park with visitors gives a sense of the attitude. Describing the trip, he said, laughing, that mountain goat stew, bighorn ragout, and "marmot au park granée" were on the menu. "There was a lot of wildlife [in the old days]. You saw a lot more than you do right now! So, we wouldn't care too much."[45] With a shocking matter-of-factness, Ruhle even told of a time that a government hunter from Columbia Falls was brought in, in order to provide more deer for people to enjoy: "In the years [19]24 and [19]25 we got rid of 450 mountain lions, and still the ecology and the environment of the park seemed to persist. The amazing thing of it was that the estimate always was that one lion would take 50 deer a year. Well, now you multiply that 450 by 50 and you come to a rather astounding figure, especially when you consider that in those days the most deer estimated in the park was around 3,000. So we had to assume that the mountain lions would come in from the outside ... this was the park in the old days!"

Moreover, white settlers inside the park were permitted to hunt on the land, and whitefish was badly over-fished within Glacier to supply the park's Louis Hill Hotel restaurant. The latter practice was banned in the early 1940s on ecological grounds.[46]

One result of this wildlife management strategy was that wolf had disappeared from the park by 1945, while the deer population far exceeded its carrying capacity.[47] Winter feeding and predator extermination were eventually stopped, though the policy clearly contributed to Glacier's reputation as a wildlife refuge.

Blackfeet resistance began soon after the park's designation. As Ruhle himself described it, park management was having "a lot of trouble with the Indians. The Indians claimed that they were being deprived of their rights by the Park Service. They said that the whole eastern half of the Park was once part of the Reservation ... The Indians claimed that they reserved the right to hunt, fish, and cut timber in that land."[48] The Blackfeet Tribal Council wrote its first protest letter to the senator of Montana in 1915 demanding recognition of traditional hunting and use rights in the eastern half of Glacier National Park. It was ignored. A petition in 1924, including many pages of signatures and more letters through 1928, were also ignored. Park management adopted a "siege mentality." A 1935 court ruling brought legal clarity to the issue, concluding that the Glacier National Park Law recognized no Blackfeet rights within the park, that these rights had, in fact, been extinguished by the 1910 law.[49] (This ruling was a sign of the times. During the 1930s, fifteen national parks were created on Indian reservations throughout the United States.)[50] In 1954 and 1966, the Blackfeet brought two more legal cases against the US government over their 1895 usufruct rights, but lost both. Again, in the 1980s, tensions heightened to 1930 levels as a new generation of Blackfeet came of age.

In recent years, Glacier park management's approach both to wildlife management and to Blackfeet culture has reportedly softened. But conflicts over hunting, fishing, and timber use as well as the management of sacred sites have persisted. And the basic disagreements about legal rights remain.

Commenting on these unresolved dilemmas, Mark David Spence writes that Americans saw Glacier as a "vestige of the 'virgin' continent ... [It] presented a fantasy realm where individual Americans could play out little frontier dramas and, like their European forebears, reinvigorate their lives through contact with the essential elements of the American wilderness."[51]

Vicky Santana, a Blackfeet woman who danced in Glacier's hotels as a young girl, says the park resulted from fraud. She says the early park proponent George Bird Grinnell, who facilitated the sell-off of the Indians' land, died insane, "partly driven crazy" with guilt over stealing the Blackfeet lands.[52]

Fathers of Thai Conservation

> *Surgery! Remove the cancer! —That is, the old corrupt foresters. It is like removing the rotten fish so that other fish in the basket and new ones that may be caught don't also become rotten . . . The Minister knows well who these rotten fish are.*
>
> Boonsong Lekagul's advice to the minister of agriculture, 1958[53]

No agriculture minister or any other Thai politician ever acted on Dr. Boonsong's foregoing advice. On the matter of national parks, however, Boonsong succeeded virtually single-handedly in pushing for the passage of new laws and introducing a new form of land management into the Thai forestry bureaucracy. While national parks and wildlife sanctuaries were perhaps the best anyone could expect to achieve at the time, they are from today's vantage point relics of the Cold War, militaristic in spirit and, being based closely on the American conservation model, inherently elitist and even racist. They are arrangements that are designed to favor the tourism industry over the interests of local people and, as such, lend themselves to intractable social conflicts. National parks empower— indeed they oblige—foresters to relocate forest dwellers in the name of conservation, and give local people no real choice but to fight back. "Both the state and the people are victims of an unjust legal structure," commented former Prime Minister Anand Panyarachun. "We simply can't go on like this."[54] Against a backdrop of continuing unresolved social conflicts in and around protected areas, this chapter looks back at two men who played key roles in bringing national parks to Thailand.

❖ ❖ ❖

When the intrepid medical doctor Boonsong Lekagul requested a personal audience with Field Marshal Sarit Thanarat in 1959 to plead the case of Thai wildlife, he knew he was taking his life in his hands. Upon seizing power the previous year, Sarit had torn up the constitution, declared martial law, and arrested hundreds of his critics in the media and opposition parties. Before the meeting, Boonsong bid a precautionary farewell to his wife and five children lest he never return.[55] Fortunately for the Lekagul family, Sarit looked favorably on Boonsong's proposal to create a system of national parks in Thailand based on the American model. Forest conservation was thus added to the Royal Forestry Department's logging mandate—not on the initiative of foresters, but on the whim of a dictator.

This was the height of the Cold War, the dawn of the development era. Following the US State Department's recognition of the new Sarit government, the World Bank made its first mission to Thailand. American economic and military aid flowed into the country a hundred times faster than all other bilateral aid combined,[56] and there was a sudden explosion in the construction of roads (mostly for military purposes), dams, and other infrastructure, all of which accelerated forest clearance and colonization. Starting in the 1960s in the northeast, the government actively promoted upland cash crops, first cassava and then Guatemalan hybrid maize. The two crops were in high international demand and tended to degrade the soil, two factors that contributed to the rapid conversion of forests to agricultural land.[57]

As Thailand emerged as a front-line state used by the United States as a base to launch its war against communism in Indochina, the Thai military—massively supported with US aid—waged war on communist insurgents staked out in Thai forests. A key military strategy for taking back territory from the rebels was road building and encouraging colonists to clear and settle the newly accessible forest, particularly in the north and northeast that border Indochina. These politically inspired military settlements could be very large. At Khao Kho in Phetchabun Province, the site of a major standoff between the army and insurgents in the late 1970s, the army created thirty-five villages on land carved out of the forest, each with around fifty families. As Pasuk Phongpaichit and

Chris Baker put it: "The insurgency was rolled back under a cloud of wood-smoke."[58] This tactic also facilitated lucrative cooperation between the military and loggers. It will never be known how much timber was cut under these conditions. Clearly though, foresters who were not directly involved in such deals could do little to control them.

Added to this, the European timber companies were facing the expiry of logging concessions that had ensured their monopoly over Thai teak for six decades. They had concentrated their efforts on what foresters call "creaming," logging the best, largest, oldest teak trees. Teak production reached unprecedented levels as firms rushed to take out as much timber as possible by the 1960 deadline. They were, of course, then replaced by local entrepreneurs and state agencies, which meant, in effect, that more players were cutting down more trees of smaller sizes. In 1960, when the Forestry Industry Organization took over three quarters of the country's teak concessions, the average size of harvested logs was about a third of what it had been thirty years before. The absence of the largest trees further facilitated the conversion of forest to agricultural land. Documented illegal logging of teak, moreover, ran at almost 50 percent of total production in this period.[59] Forest cover decreased rapidly, dropping from 60 percent in 1954 to 43 percent just twenty years later, an average loss of 4,500 sq km a year.[60]

Against this background, the fairy tale notion of setting aside areas of forest as nature preserves contradicted the dominant trends of a time that has otherwise been aptly called Thailand's "Forest Invasion" period.[61]

Boonsong Lekagul was desperate about the situation and for more than a decade had been using every possible means to awaken the Thai public to the implications of forest loss for wildlife. It is probably no exaggeration to say that by 1959 he was already something of a living legend; he is certainly one of the world's pioneer environmentalists of the post-World War II period.

Though trained as a medical doctor, Boonsong's great passions were forests and wildlife. His *Niyom Phrai Samakhom* (Association for the Conservation of Wildlife), established in 1947, was Asia's first environmental organization. In 1958 he published *Hak pa mai yang yu yang yuen yong* (If the Forest is to Last Forever), which is a detailed critique of Thai forestry laws, policies, and practices, along with his proposals for reform. In addition, he reportedly wrote or translated some thirty-five books in English and Thai, including three classic volumes on wildlife, *A Guide to the Birds of Thailand* (with Kitti Thonglongya, 1968), *The Mammals of Thailand* (with Jeff McNeely, 1977) and *A Field Guide to the Butterflies of Thailand* (with K. Askins, J. Naphitabhata, and A. Samruadhit, 1977), illustrated with his own paintings, photographs, and sketches. He trained and inspired generations of young people through his public lectures, his countless guided expeditions to forest areas, the journals he founded, and his radio talk shows. Notable among his awards are the Order of the White Elephant from H.M. King Bhumibol and the World Wildlife Fund-administered Getty Prize.

Ironically, Boonsong's commitment to wildlife conservation grew out of his vast experience as a game hunter. Like the European and American safari hunters in Africa who did not find it inconsistent to kill the thing they loved, Boonsong appears never to have felt regret about the great numbers of animals he killed, arguing that his aim was never profit but scientific research and education. Boonsong's enormous collection of antlers and stuffed and preserved specimens was meant to form the core of a Thai natural history museum. (That project was never realized.) Boonsong's career as a hunter begs comparison with US president Theodore Roosevelt, whose notorious African safari of 1909—described even by his fellow hunters as outright "slaughter"— was ostensibly undertaken to collect specimens for the Smithsonian Institute and the American Museum of Natural History.[62] But Boonsong's style of hunting was different. The American president had traveled Maharaja-style in a mile-long caravan with servants and cooks. Yet it is said that as a medical student in the early 1930s, Boonsong would travel on weekends to roadless forest areas around Bangkok, hiring local hunters as guides and renting or buying villagers' carts to carry his equipment. They would track animals by foot, an activity that was physically strenuously and brought the young doctor in close touch with the forest. He came to admire his hunter guides whose "code of honor" demanded that they minimize the animals' suffering, avoid killing females and their young, and hunt only for their families' subsistence and not for sport or profit.[63]

Boonsong had a special disdain for the style of game hunting that proliferated after the Second World War where rowdy men in jeeps armed with automatic military hardware drove into the forest and shot at anything on four legs. This he viewed as lazy, irresponsible, immoral, and indiscriminate killing by hooligans, and eventually contributed to his decision to stop hunting entirely in the late 1940s.[64]

Boonsong was under no illusions about the inherent risks of putting wildlife protection in the care of soldiers. The year before his historic meeting with Field Marshal Sarit, he wrote: "Since soldiers carry weapons they carry power in their hands, political power . . . Therefore if they are not educated about conservation . . . [they] could use military force or their political influence to go in and destroy forest and wildlife, and claim forest land for themselves, and government officials would not dare to stop them."[65] Probably stretching the truth to make a point that was relevant to his own country, he noted that "in other countries all around the world, conservation has top priority and natural resource education is given at all levels . . . especially in army, navy, and police schools and officer colleges."[66]

But if Boonsong was skeptical about the military's capacity to exercise self-restraint in matters of the forest, he had even less faith in Thai forestry authorities. His frequent visits to forests throughout the country gave him the opportunity to talk to foresters of all ages. He wrote that forestry students started out as "good and pure-hearted" men who looked forward to a bright future of earning a living by honest means. He blamed their training for failing to teach them the one thing that would make them good foresters: a love of the forest and the living creatures within it. Instead, forestry education turned them into morally corrupt officials who did not perform their duties and instead spent most of their time in "unwholesome environments like billiard halls," he wrote. "I feel that the Forest College is not able to teach . . . its students to love and be interested in the natural forest, which should be an essential foundation for a good forester . . . This is nothing for Kasetsart University to be proud of."[67]

It is not hard to understand that Boonsong had difficulty persuading the Royal Forestry Department to embrace conservation—a concept that was at once fundamentally at odds with Thai forestry's original mandate and would have demanded that foresters forgo logging revenues in large, extended areas. In any case, to create national parks new laws had to be passed. As Thailand was a military dictatorship with no democratic channels for influencing policy makers, Boonsong had no option but to go straight to the top. But he needed allies. And given the overwhelming influence of American thinking, financial support, and technical advisors at the time, it made sense to bring in an American advisor to support his case.

If there was a pull from the Thai side, there was also a push from the American side. George Ruhle's arrival on the scene in 1959 coincided with a shift in the focus of the American conservation movement from an exclusive concern with its own wilderness areas to the internationalization of the national parks model. According to one wilderness advocate, the United States had a "national parks system which [was] the unrivalled adornment of the hemisphere."[68] Organizations like the Sierra Club, founded by John Muir, the International Union for the Conservation of Nature (IUCN) and the World Wildlife Fund (WWF) spearheaded the drive to export this apparent American success story. As Ruhle himself explained, the International Commission on National Parks of the IUCN had been "interested since its establishment in furnishing technical aid to countries in need of such assistance."[69] National parks were defined by IUCN as large areas "not materially altered by human exploitation and occupation."[70] A Sierra Club executive even called wilderness "a powerful diplomatic weapon" that would help overcome the hostility that some countries harbored against the United States and would counteract the belief that America was merely an "industrial nation with a dollar sign for a heart."[71]

For these prophets of conservation, national parks represented the pinnacle of achievement for a civilized nation, evidence of a country's moral standing. It is as if the world was divided into two categories: those that are rich and sophisticated enough to appreciate wilderness, and those too poor and backward to do so. American wilderness historian Roderick Nash typifies this attitude in his description of nature appreciation as a "full stomach phenomenon," confined to the rich, urban, and sophisticated. A society must become technological, urban, and crowded before a need for wild nature makes economic and intellectual sense, he wrote.[72] An Australian deep ecologist declared even more bluntly: "We are the only countries in the world capable of doing anything about [the forest crisis] ... people in Brazil or Indonesia are not in a position to protect their forests."[73] Judging from his own writing, Boonsong did not fall for such nonsense. Still, it is probably safe to assume that during his 1959 meeting with the pro-American Sarit, he argued that establishing national parks would contribute to Thailand's reputation as a civilized, sophisticated society.

As for George Ruhle, he is more than just a passing figure in the great project of exporting American-style conservation. Ruhle had, in the words of a US National Parks Service officer who worked with him for many years, "an illustrious career" and was "the best America had to offer as a consultant on conservation at the time."[74] Not only did he work for thirteen years as chief naturalist at Glacier National Park, he had also worked at both Yosemite and Yellowstone. It was, in fact, Stephen Mather, the first director of the US Parks Service, who personally encouraged Ruhle to join the Glacier staff on a permanent basis.[75] Ruhle represented American conservation ideas in their original form.

Ruhle did not only travel to Thailand. After leaving Glacier National Park, he advised on conservation issues in Taiwan, Korea, Japan, Argentina, Guatemala, Indonesia, India, and East Africa.[76] Upon returning to the United States from his period in Thailand, Ruhle set up the Division of International Parks Affairs within the US Parks Service, with the personal support from then Secretary of the Interior Stewart Udall who served under President John F. Kennedy.[77]

Ruhle had no wife or children, and devoted much of his time and personal resources to the cause of conservation. He received no salary for his work in Thailand, and even paid for part of his own travel. Formally, Ruhle represented the IUCN and the American Committee for International Wild Life Protection. His task was to assist the Royal Forestry Department and Dr. Boonsong's Association for the Conservation of Wildlife with the formation of a national parks system. Boonsong is described in Ruhle's report from the trip as a "prime mover in initiating the project."[78]

During his months in Thailand, Ruhle traveled to eleven potential national park sites, including Khao Yai, Doi Suthep, Doi Pui, Doi Inthanon, and Phu Kradueng. Though none of these places "necessarily [complied] with the lofty standards set for national parks" in his opinion, they did have national significance that made them worthy of preservation. "The fact that the need is so urgent, the continuing loss so great, and the progress so desperately slow is no reason for giving up and doing nothing at all," Ruhle stated.

The cure for Thailand's ills was the creation of a system of national parks that would preserve forests as "inviolate watersheds." As he wrote in his "Advisory Report on a National Park System for Thailand 1959–60," his assistance was "requested in preparing legislation on conservation while I was in the country. Initial drafts of laws were reviewed and edited, culminating in the passage in 1960 of new, stronger forestry ordinances and a game law which gives a measure of protection to many species ... In addition ... a proposed national parks law for the establishment and operation of parks and reserves was passed in principle on January 14, 1961." Ruhle also hoped that an interpretive program "patterned after that in the United States national parks [could] eventually be established."

He was appalled to discover that, even at the highest levels, Thai people seemed to think of national parks as "amusement parks, or resorts for drawing huge money-spending crowds." Moreover, "few Thai recognize that native plant and animal life ... are precious natural resources ... Like other peoples of Southeast Asia, the Thai are squandering their natural resources." As for the causes of Thailand's forest demise, Ruhle had no doubts. He did not mention road building or cash crops, military forest colonization, legal or illegal logging, rapacious cutting by the foreign timber companies as they rushed to log out as much as they could before their licenses expired. Nor did he predict the potentially devastating impact of Thailand's first big dam, Yanhi (later renamed Bhumibol), which was by then under construction. Rather, the country's biggest problem, according to Ruhle, was ethnic hill people. "What to do about the unassimilated non-Thai tribes of the forested highlands is a pressing problem. It is folly to overlook their existence, for their mode of living and shifting cultivation destroys critical watershed and poses a serious threat to water and soil resources of the lowlands upon which the nation's future welfare depends," he wrote. Without offering any evidence, Ruhle further asserted that these "non-Thai" peoples were causing *more damage* than the loggers. "The *kaingin* [slash-and-burn] farmer annually destroys more valuable timber than the lumberman turns into lumber," he wrote.[79]

The 1960 Wildlife Conservation Act and the 1961 National Parks Law, which George Ruhle helped to draft, allowed for the creation of two main categories of protected areas in Thailand: wildlife sanctuaries, where human presence is restricted mainly to scientific research; and national parks, where tourism in designated areas is encouraged. Thai foresters did not ask for conservation to be added to their logging mandate, nor were they trained in conservation. Conservation was, rather, thrust upon them. Two new divisions of the Royal Forestry Department—the Wildlife Conservation Division and the National Parks Division—were created to manage these new categories of forest land. In effect, with the passing of these laws, foresters were required to perform contradictory functions, the management of timber extraction, on the one hand and the guarding of "virgin" forest, on the other. (These tasks were presumably to be carried out on different areas of forest.)

Problems associated with giving responsibility for protected areas to a forestry department accustomed to viewing the forest as a source of revenue became apparent right from the beginning. The Committee for the Improvement of Khao Yai was formed in 1960 in anticipation of the new law. In recognition of his key role in bringing national parks to Thailand, Dr. Boonsong was invited to join the committee along with a number of directors general and the minister of agriculture. That committee met only once on February 17, 1960, to discuss the request of an Italian firm to build a road into Khao Yai and to lease 5,000 rai of land in the middle of the park for twenty-five years to build a hotel and casino. At the meeting, Boonsong—speaking his mind as usual—commented that anyone willing to sell Khao Yai to these foreigners should be incarcerated or used as target practice! That committee was dissolved, and Boonsong was not invited to sit on the second park management committee that was later formed. Two years on, it approved the golf course in the middle of Khao Yai and Boonsong could do nothing to stop it.[80]

Boonsong is an interesting figure because of the ways in which his thinking changed over the years. His transformation from famous hunter to famous conservationist has already been noted. Nadda Sriyapai, the doctor's friend and protégé, comments that on the matter of hydroelectric dams, Boonsong's ideas also went through a complete turnaround. In the early 1960s, he was a great believer that dams would bring countless benefits to Thailand. About the World Bank-financed Bhumibol hydropower dam, which was commissioned in 1964, he wrote: "People will have cheap electricity for [their daily needs], reducing their use of fuel wood and [thereby] lessening the devastation of forests." Other advantages included irrigation for agriculture, improved water transport, protection of national parks, and prevention of saltwater intrusion in Bangkok and Thonburi. Moreover, he imagined that on the "two tiny lakes" formed behind the dam there would be flocks of ducks, as well as "tens of thousands of water birds of all types."[81] The reality of the devastating ecological and social impacts that the Bhumibol and other dams have caused in Thailand resulted in Boonsong revising his opinion. By the time the first plans to build Nam Chon became public in the 1970s, he had become a vociferous critic of hydro dams.

In a poignant footnote to Thai environmental history, Nadda recounts that immediately following the announcement of the final decision to shelve the Nam Chon dam project he went to Boonsong's house to deliver the news. By 1988 Boonsong was an old man, confined to a wheel chair and paralyzed due to a stroke. When Nadda told him that Nam Chon had finally been canceled following massive public protests, Boonsong did not speak or even smile. His only reaction was a slight nod of his head, the only movement of which he was capable, indicating that he understood.

Conservation Unlimited

A set of rules used in one physical environment can have vastly different consequences when used in a different physical environment.

Elinor Ostrom, *Governing the Commons*[82]

The logging ban and revocation of the *Khor Chor Kor* military forestry resettlement scheme threw the Royal Forestry Department into a crisis. Legal timber extraction was no longer possible. And reforestation—defined as large-scale plantations of exotics, which had been a top priority for the Thai forestry establishment since the early 1980s—came under frontal attack from the popular resistance movement. Recall that 1992, the year *Khor Chor Kor* was suspended after popular revolt, coincided with the Earth Summit, and Thai public concern about the environment was in the air. RFD responded by remaking its image, shifting its primary focus from logging and eucalyptus plantations to forest conservation.

From under-funded, low profile divisions that had limped through the decades following Dr. Boonsong's dramatic intervention in 1958, the importance of the conservation wing grew dramatically. Budgets and staff were increased, and there was a major push to accelerate the designation of protected areas. From 1989 to 1995, about 15,000 sq km of forest were added to the protected areas system, extending it from 9 to 12 percent of the country's surface area.[83] The process was pushed along after the creation of the National Parks Department under the new Environment Ministry in 2002, so that by 2008, conservation areas covered 96,000 sq km of forest, or almost 19 percent of the country (see table 3).[84] The grandiose project of expanding the gazettement of Thai conservation territory is still not finished. According to Director General of the National Parks Department Chalermsak Vanichsombat, the goal is to declare forty more national parks and three new wildlife sanctuaries. When this process is complete, terrestrial conservation areas will cover 117,000 sq km, a formidable 23 percent of Thailand's surface area.[85] Curiously, according to UNEP global statistics, the same proportion of territory has been set aside for conservation purposes in the United States.[86]

Table 3 Thailand's protected areas, 2007

Current protected areas	No.	Rai (millions)	Sq km	Total (%)
National parks	108	34.20	54,720	
Wildlife sanctuaries	57	22.63	36,208	
Non-hunting areas	60	3.27	5,232	
Total existing protected areas	225	60.10	96,160	18.7
Preparing for gazettement				
New nation parks	40	11.64	18,654	
New wildlife sanctuaries	3	0.44	704	
Annexes to national parks	3	0.19	304	
Annexes to wildlife sanctuaries	12	0.92	1,472	
Total areas under preparation	58	13.19	21,134	4.1
Grand total	283	73.29	117,294*	22.8

*51,800 sq km (32.39 million rai) designated as Class 1A and Class 1B. Watersheds are not represented in these figures since they are not formally defined as protected areas.

Source: Interview with National Parks, Wildlife and Plant Conservation Department, Bangkok, January 3, 2008.

The supremacy of the conservation agenda is at first perplexing, given the traditional antagonism within the bureaucracy between production and conservation forestry, and the dominance of the commercial production wing for much of the previous century. Indeed, the new agenda involves a colossal reversal of mandates; a radical shift in the goal of Thai state forestry—from an agency tasked with overseeing the extraction and processing of timber from forests to one responsible for protecting forests from timber extraction and other forms of encroachment. On the surface, the Forestry Department's new identity as conservation advocate marked an abrupt break with the past; its *raison d'être* switched from forest exploiter to forest protector. But this apparent reversal obscures continuity on many levels.[87]

First and foremost, the recent intense focus on conservation reflects continuity with the past in that it is, in fact, a direct response to the failure of Thai state forestry. Logging was not practiced sustainably. Replanting, either by companies or by foresters, was never taken seriously. Corruption was so rampant as to be institutionalized. Notwithstanding the sustainable forestry rhetoric, Thai forests were treated like a non-renewable resource and more than two-thirds of the country's forest cover disappeared in the course of the twentieth century. The logging ban, issued in response to public outcry, triggered the change of outward focus. But basic inner structures remained intact. Two major scandals that occurred in the years following the ban exposed how little had really changed.[88] In the Tha Chana case in 1996, massive illegal logging in the southern province of Surat Thani resulted in the arrests of the RFD director general, the deputy minister of agriculture, and a dozen other high-ranking forestry officials found to be deeply involved in an illicit timber network that also included politicians and business people. Many believed that Tha Chana was just the tip of the iceberg. Two years later, large-scale illegal logging inside the Salween National Park along the Burmese border was exposed. Once again, leading forestry officials, from the heads of the National Parks and Wildlife Sanctuary divisions up to the director general himself, were implicated in the criminal activities.

These scandals failed to provoke a fundamental questioning of Thai state forestry ideology though. While the mandate had shifted, three important aspects of the bureaucracy remained unchanged: the basic focus on forest as a source of income generation, the state monopoly over forest areas, and the legal framework that continues to withstand far-reaching reform.

Income generation

A key element in the Thai state forestry ideology is the view of forests primarily in terms of their commercial worth. Recognition of forests' ecological dimensions has had remarkably little impact on Thai state forestry thinking, policy, or law. Thus with the expansion of national parks, areas have been designated more for their scenic value and potential as tourism destinations than for their actual ecological values. This can be seen in the overwhelming emphasis on gazetting national parks. The new areas being prepared for gazettement as national parks where tourism is permitted outstrips areas being prepared as new wildlife sanctuaries where tourism is not permitted ten to one. Moreover, a majority of Thai parks and sanctuaries have been selected in areas with mega-fauna, large mountainous forests, waterfalls, beaches, and islands—tourist attractions, in other words—while key ecosystems such as mangrove forests and lowland riverine forests and wetlands were at first completely ignored because they were deemed to be less appealing. Pinkaew Laungaramsri says profit making has been the main goal of park management and largely determines how scarce resources are allocated. "Commodifying nature through the promotion of 'natural wonders' has been the key marketing strategy of the tourism industry," she writes.[89] While the discourse has focused on "pristine," "core," "unspoiled" areas and biodiversity preservation, the reality has been rampant hotel and resort development, golf courses, and the like inside national parks. The first golf course built inside Thailand's first national park was only the beginning. The story of the eight dead baby elephants found inside Khao Yai illustrates the precarious situation of flora and fauna inside Thai protected areas as tourism development proliferates.

It happened during the rainy season some years ago in the heart of Khao Yai National Park. Eight baby elephants were found dead at the bottom of a waterfall. Apparently, their path led straight to the edge of a cliff, and they were washed over the edge by the surging current. On the opposite side of the country, then Chief of Thung Yai Naresuan Wildlife Sanctuary Veerawat Dheeraprasart heard the news by radio at his headquarters. The announcer reported that plans were underway to build a wall or some sort of obstacle to steer the animals away from the dangerous cliff, and so avoid further accidents. The name of the area, "Hell's Ravine" (*Haeo Narok*), was already well known as plans to build a hydroelectric dam at the site had only recently been shelved.

Veerawat sensed that something was not right. "It seemed to me that the government was as usual trying to solve a problem by dealing with the symptom instead of the cause," he recounted later. He asked the opinion of Neah Teuh, a Karen elder of Krueng Bo village inside Thung Yai.

Neah Teuh's immediate response was that it was "impossible" for an elephant path to lead directly over a dangerous precipice. An adult female would always follow the safest possible route through the forest, especially when accompanied by her young. The elephants are like the forest Highway Department, clearing a path that other animals (and humans) in the forest community rely on to move about, he said. Neah Teuh offered three hypotheses for what might have caused the deaths in Khao Yai.

The first was that the elephants were frightened by something that had caused them to bolt—perhaps the sound of cars beeping their horns on the road, which passes quite close to the waterfall. Or perhaps the sound of someone shouting and chasing them, or perhaps the sound of guns. The park rangers would know the movements of elephant herds well. Had anyone checked if there had been hunting in the park? The second hypothesis had to do with the animals' feeding habits. Perhaps all the elephants' safe feeding grounds had been destroyed or were disturbed by human presence in the park—by the hotel, the golf course, and the road as well as by the thousands of visitors who go to Khao Yai each year. Perhaps the animals were feeding near the waterfall for lack of other safer sources of food. Neah Teuh compared them to a hungry man, so desperate for food that he is no longer thinking clearly and is willing even to risk his life for a bit of something to eat. Perhaps, he said, the elephants were starving.

The third hypothesis struck Veerawat as the most interesting because it reflects the unique worldview of the Karen, a people known for their knowledge of elephant behavior. Maybe the elephants in Khao Yai had lost their instincts. After living for so many years near the trappings of city people, perhaps these animals had become accustomed to exhaust fumes and garbage and headlights and golf balls, all of which had impaired the instincts they need to survive in the forest. Perhaps,

Neah Teuh said in an unmistakable reference to the Karen peoples' own predicament, the old ones that had this knowledge had died off, and the younger generations had lost their "forest sense."

Another more recent, disheartening example is the plan to create a highway inside the newly declared World Heritage Site, Dong Phayayen-Khao Yai Forest Complex, which encompasses Khao Yai National Park and surrounding areas. According to UNESCO, the area contains more than 800 fauna species, including 112 species of mammals, 392 species of birds, and 200 reptiles and amphibians. It is deemed to be internationally important for the conservation of globally threatened and endangered mammal, bird, and reptile species that are recognized as being of outstanding universal value.[90] But the 2005 declaration had strings attached; the UNESCO World Heritage Committee warned Thailand at the outset that the integrity of the site was at risk due to ecological fragmentation of the forest. Of special concern to the committee was the Highway Department's plan to expand an existing route through the forest complex into a four-lane road, Highway 304. The existing, smaller road currently divides Khao Yai and the adjacent Thap Lan National Park. It is the main route for transporting industrial goods and commodities between the northeast and the deep-sea port at Laem Chabang in Chon Buri Province, and the expansion project of the road is viewed as having high strategic economic importance.

The National Parks Department administration claims to be powerless in this situation. Former department director Damrong Pidej said he was opposed to the project because it would "tear apart the forest complex and seriously damage wildlife habitat."[91] In addition to blocking migration routes, many wild animals had been struck by cars and killed on this road, he said. Widening the road would only worsen the situation. Damrong claimed to have tried to convince the Highway Department to revise the project, but they had refused. Certainly, Thai conservation law was not on Damrong's side. Environmental impact assessments are not even required for road expansion within protected areas. Nevertheless, the Highway Department has been forced to conduct an EIA—not by Thai forest conservation authorities, but by the World Heritage Committee, which at its annual meeting in July

2007 issued this as a demand. The committee noted that an enlarged Highway 304 would exacerbate the problem of ecological fragmentation that already plagues the forest complex and gave Thailand one year to produce an EIA. But no environmental impact assessment was submitted to the July 2008 meeting of the World Heritage Committee, and so Thailand was given two more years to submit a report.

If Thailand cannot demonstrate that it is addressing the fragmentation issue, however, the Khao Yai complex could be put on the "endangered heritage sites" list. Should Thailand push ahead with the expansion, the site could well be deleted from the World Heritage list. (To his credit, Damrong's successor Chalermsak Vanichsombat has proposed to the cabinet a new regulation that would forbid road construction and expansion inside the country's protected areas except, he notes, for routes used by park management or for tourism purposes.)[92]

One proposed solution that would accommodate traffic while lessening impact on wildlife is to put the two-kilometer stretch of road underground. But this will significantly increase the cost of the road, and the Highway Department doesn't want to pay. Ultimately, the price of such a tunnel will have to be weighed against the loss of face that the Thai government must endure if this World Heritage listing is revoked.

State monopoly control

The asymmetry of the state's approach to conservation is revealed when we compare the "powerlessness" of forestry authorities in the face of highway construction inside a prestigious conservation area, on one side, with the fervor foresters have often shown in evicting local people from parks and sanctuaries, on the other.

For a department that had once, at least on paper, held jurisdiction over almost half the surface area of Thailand, the cancellation of timber concessions led to a dramatic erosion of territory under its control. The Agriculture Ministry's forest lands survey in 1992 defined over 70,000 sq km within legally designated forest as being either already converted to farmland or too degraded to be considered for conservation purposes. In one fell swoop, this huge area (13 percent of the country) was taken away from the old RFD. Given this harsh reality, the mantel of forest conservation became a lifeline for the weakened forestry bureaucracy, which embraced the new agenda as a way of restoring its sphere of influence and its monopoly control over forest areas. The frenzied expansion of national parks and wildlife sanctuaries helped the forestry establishment regain both lost authority and lost territory. Enlargement, and ultimate dominance, of the conservation mandate was driven forward by the approach to conservation that had been imported from the United States, one that echoed the old colonial scientific forestry. Both are predicated on absolute state jurisdiction over forest. But the newly gazetted lands were of course not uninhabited and, inevitably, new conservation boundaries overlapped with forests where people lived—enclosing some existing settlements and cutting others off from forests they were accustomed to using and managing. Many communities were faced with the ironic injustice of losing access to forests they had defended against commercial or state projects as these new conservation areas were declared off-grounds to people. As the forestry bureaucracy redefined itself as the champion of forest protection, popular resistance grew and the movement to democratize forest management gained momentum.

Recall that American national parks, on which Thai protected areas are modeled, were based on the idea of un-peopled wilderness. As we have seen, the "emptiness" of these places and their characterization as "virgin" nature was predicated on the removal of people who had called those same places home. In Thailand, where forest use and management by communities have a long and complex history, this wilderness myth was translated into laws prohibiting human habitation inside conservation areas. As George Ruhle had done five decades earlier, Plodprasob Suraswadi, the influential director general of forestry in the post-logging ban period, identified shifting cultivation as the main culprit in deforestation. In Ruhle's sublime wilderness, park authorities had been "in a near state of war" with Indians who claimed a large part of Glacier park to be their land.

Both Yosemite and Yellowstone were run by soldiers for half a century because of ongoing conflicts with Indians. From this perspective, the militarization of Thai state conservation that intensified under Plodprasob at the close of the twentieth century, far from being a perversion, was true to its American ideological roots. The cover of the Thai forestry magazine, *Wanasan*, in its first issue in the year 2000, reflected the mood of Thai state forestry. The headline, "Forest Fire Protection: Fast and Successful," was accompanied by a photograph of Plodprasob armed with a gun and a knife, along with his men who were standing like soldiers prepared for combat.[93] During the five years Plodprasob headed the Forestry Department, foresters, army, and police cooperated actively in a campaign against the so-called "hill tribe" people in protected areas. Villages of Hmong, Karen, Akha, and other groups were raided in Chiang Mai, in Nan, in Kanchanaburi; people were intimidated, orchards were destroyed, mass arrests were made, and evictions were carried out.[94]

In retrospect, if it had not been for the failed *Khor Chor Kor* scheme, the level of repression might well have been even worse. One might have expected forestry authorities and the military under Plodprasob's leadership to come up with a consolidated mass resettlement scheme to remove all forest dwellers from conservation zones. After all, while the military forest resettlement plan had brazenly aimed to move some 6 million people from their homes to make way for tree plantations, the number of people living in protected areas is likely in the order of 2 million. But *Khor Chor Kor*'s demise revealed the limits of public tolerance. In spite of the frantic, racist rhetoric, there was never a single systematic hill tribe resettlement plan on the scale and level of detail of *Khor Chor Kor*. It seems safe to say that such a plan, even given the resonance of anti-hill tribe discourse with the general public, would not have been politically acceptable. Instead, therefore, ethnic minorities, who were less well organized and politically more vulnerable than many of the lowlanders targeted by *Khor Chor Kor*, were subject to a sporadic, selective campaign of repression, one which Larry Lohmann has referred to aptly and disturbingly as "forest cleansing."[95]

Legal structures

Before turning from this discussion to the question of the future of Thai state forestry, it is important to bear in mind, finally, that the switch from a timber-based mandate to a conservation mandate has occurred within a virtually unchanged legal framework. With the crucial exception of the Community Forest Bill, whose origins are completely different than all other Thai forestry legislation, the laws that govern Thai state forestry are based on outdated, even antiquated, ideas. Nevertheless, the bureaucratic resistance to legal reform continues to lock Thai foresters into old legal structures and ways of thinking.

The 1941 Forestry Act, for example, which is still in force today, defines "forest" in terms of the amount of standing commercial timber. Logically then, according to the 1964 Forest Reserve Act, "degraded forest" is defined as land without standing commercial timber. And while the 1985 forestry policy talks about reforestation as a way of combating environmental degradation, it fails to introduce an ecological dimension into the definition of forest. Therefore, since degraded forest refers to an absence of timber, "reforestation" becomes equated with the planting of fast-growing exotics. This conceptual structure perpetuates the dichotomy between natural forest and economic forest and encourages a relentless conversion of "degraded" forest to monoculture in the name of "reforestation." Ecological restoration, conversion in the opposite direction, as it were, from degraded to more healthy forest, is virtually impossible because it has no conceptual or legal "home." This has led to the absurd situation where communities that have restored degraded forests have lost access to these areas as they are included into newly declared parks and sanctuaries.

REINVENTING THAI FORESTRY

We are still slaves to the borders that we draw, even though people were there before. But the borders can't conserve the forest; it is only the people who can.

Anan Kanjanapan, anthropologist[1]

What would a Thai forestry based on Thai forests look like?

For over a century, Thai state forestry has had an illusory quality to it, being founded not on existing biological and cultural realities, but on ideas born in other geographic places and times. State forestry ideology in Thailand consists of a gospel in three parts—scientific forestry, industrial tree plantations, and wilderness-based conservation. Each has been adopted unquestioningly, together forming the intellectual foundations of a centralized, authoritarian bureaucracy that demands obedience and allegiance, and discourages critical questioning. Forestry professor Somsak Sukwong commented many years ago that there is no such thing as tropical forestry because forestry as it is practiced in tropical countries like Thailand does not take biological diversity and the deep connections between human communities and forest ecosystems as its starting points. Thai forestry has been a mediocre copy of Western approaches to forest management, ignoring the qualities that make Thailand's forests unique, a hall of mirrors reflecting imported and increasingly outdated land management ideals.

While criticism of Thai state forestry has raged for two decades, the forestry establishment has managed to insulate itself fairly successfully. In spite of the collapse of its original *raison d'être*, timber production, and the shift of focus to conservation, the same set of laws are in place and the same basic set of beliefs governs the way Thai foresters view their mandate. Somsak, a former dean of the Kasetsart University forestry faculty and the founding director of the Regional Community Forestry Training Center (RECOFTC), recently reflected on the continued resistance to change within the forestry bureaucracy: "The forestry hierarchy still exerts a strong influence over its people. This causes lower-ranking foresters not to dare to express their opinions," he says.[2] Nevertheless, he observes that whereas in the past virtually all graduates of the forestry faculty went straight into the forestry bureaucracy, the younger generation is now joining NGOs, research institutions, companies, and universities. "This will eventually change the way of thinking," he says.

Indeed, it was debate and pressure from non-forestry institutions and farmers' organizations that resulted finally in a Community Forest Bill being passed in November 2007. But opinions are divided among Thai state forestry's harshest critics about strategies for change. Witoon Permpongsacharoen and Veerawat Dheeraprasart of Foundation for Ecological Recovery were among the first to put community forestry on the national agenda in the late 1980s. Asked to reflect on twenty years of struggle to force a community forest bill into existence, they argue that this was not the right strategy. Given the power of the bureaucracy in Thailand, attaining a law that fully served communities' interests was and remains impossible, and NGOs and communities have put too much energy into arguing over legal texts with bureaucrats. "The strategy should have been to help expand existing community forests, to build up the movement, to put their practices down in writing, and present this to the authorities," says Veerawat. He says a law is not really necessary since the Constitution (both of 1997 and of 2007) guarantees community rights.

Chiang Mai University anthropologist Anan Kanjanapan, who has supported the process of writing and rewriting draft versions of the law, defends the effort, saying: "It is better to have a law than just having the Constitution because a constitution can be ripped up at any time. The laws remain." He acknowledges that the Community Forest Bill, even in its current form, marks a historical landmark and a major victory for the people's movement. Still there is extreme despondency about the version that was finally passed, and Anan accepts reluctantly that the struggle to improve it will continue.

Meanwhile, there are signs that even within forestry circles there is a realization of the need for a radical shift in state forestry thinking. In mid-2007, the Kasetsart Alumni Association, whose members consist of retired professionals in forestry and other life sciences, published its recommendations on Thai forestry policy reform.[3] The document, which is essentially an appeal for more democratic, ecologically based forms of forest management, warns that existing laws are so out of touch with reality that they tend to criminalize all non-state uses of forests. "Forests are so important that solving the forestry legislation crisis should be on the national agenda," it states. A major flaw in existing forestry laws is that they lack a consciousness of human rights in forest management. A community forest bill is therefore needed, the document notes, to support those communities that are managing forests in a sustainable way, and to set standards for others. In the absence of a law, enforcement is applied selectively. "The fact that 50,000 signatures could be collected for the bill proves that there is a real popular demand," it states. The Kasetsart elders recommend legal reform either through radical revision of the 1941 Forestry Law, the 1964 Forest Reserve Law, the 1960 Wildlife Conservation Act, and the 1961 National Park Law or by replacing them with a brand new law that gives people a role in managing their own community forests. Another key recommendation is to reclassify the country's forests, not in terms of bureaucratic, provincial categories, but in ecological terms according to catchment areas.

The Kasetsart Alumni document was published just months before the Community Forest Bill was finally passed. While activists have focused on its shortcomings, some foresters who have battled the system from the inside are more optimistic. Retired senior forester Sitthichai Ungphakorn, an author of the Kasetsart Alumni document, views the enactment of the Community Forest Bill as a cup half full. While working as regional forester in Chiang Rai, Sitthichai was well-known for going out of his way—even in violation of forestry law—to accommodate local people's customary forest use in a model he calls "collaborative forest management." He says he views the new law from a practitioner's standpoint and emphasizes the gains that it represents. Though it clearly has deficiencies, it is nevertheless an important step forward because of the community rights that it does recognize. The law provides something to build on, says Sitthichai, pending more far-reaching legal reforms. The current director of RECOFTC, Nepalese forester Yam Malla, also argues that an imperfect law is better than no law at all: "If you don't have a supportive policy and law, people remain vulnerable no matter how long they have managed the forests." Referring to ongoing efforts to challenge the constitutionality of the bill, he says this will only strengthen the opponents and potentially risks delaying the process for another eighteen years. "Then we are talking about generations," he says. "We don't have that much time."

This sentiment is echoed by another senior forestry veteran Komol Praekthong, who has grappled with community forestry within the forestry bureaucracy for much of his professional life. Reflecting on his own thinking process, he says: "I was educated at the forestry faculty of Kasetsart University, but what I was taught did not fit with the reality I knew from my childhood. We had our own cultural beliefs and practices about the forest. Thailand should be a world leader in the forestry field, but our thinking systems must be based on our own knowledge, instead of blindly following systems that come from Germany or America." Asked what it will take to create a Thai forestry, he replies: "This will not be possible until we accept the existence of community forestry."

Whether the current version of the Community Forest Bill survives or is scrapped in favor of the possibility of a law that grants stronger rights to communities, the challenge to state power is undeniable. Veteran journalist Sanitsuda Ekachai summarizes it this way: "Given its past century of failure, should we allow the Forestry Department alone to continue managing our fast-dwindling forests?[4] ... [The department] comes under pressure from all sides to carve up our precious forests for the benefit of big players in politics, business, the military, and other state agencies ... The department likes to project itself as a lone, victimized hero holding back forest encroachers. But the image bell rings. The record says it all. It bullies powerless villagers and buckles before political and business elites."[5]

The rethinking of forestry is well underway in many parts of the world. In temperate and boreal countries, the basis of conventional forestry approaches is being fundamentally questioned. German forestry policy has since the late 1990s been turned on its head. Siegfried Lewark, a forestry professor at the University of Freiburg—the first place in the world where forestry was taught in the eighteenth century—worked on the redesigning of the forestry curriculum. He emphasizes the need for forestry students to learn to think critically and to work cooperatively rather than simply taking orders. James Kennedy at Utah State University has worked with a group of foresters under the World Forestry Congress looking into redefining Western university forestry education in the twenty-first century. He cites the need for "more open, inclusive, adaptive models that can accommodate the diversity, complexity, and interdependence of ecological and human systems."[6] Kennedy's group pointed out that foresters have traditionally believed in "clean dimensions and boundaries," and have been inherently suspicious of complexity and diversity. Forestry in the twenty-first century will require that foresters have a "more humble and holistic" ecosystem view in which the forest ecosystem is composed of interdependent, dynamic sub-systems. Swedish forestry policy, once exclusively focused on maximizing timber production, now has protecting the environment as its primary goal. Forest biologist at Umeå University, Olle Zackrisson, warns tropical foresters of the heavy ecological costs of the kind of forestry practiced in Sweden, and of the inherent dangers of applying these techniques in the diverse forests of the tropics.

The main challenge of this "new forestry" in Western countries has been the inclusion of species other than timber trees and the "ecological services" of forests in the landscape. But in Thailand, the task is far more complex because of much higher levels of biological diversity and because of the human dimension. Thai foresters must not only broaden the scope of species to include more of the ecosystem; they must also build alliances with forest-dependent communities. This is the most difficult part of the transition since local people have been the nemesis of state forestry ideology through all its three main historical periods—the timber phase, the industrial plantations, and the current conservation era. The change will require the creation of new administrative structures, laws, and policies, and an education system that gives priority to documenting and developing existing knowledge of medicinal forest plants, food plants, and other non-timber products, traditional forest agriculture and management systems, cultural and religious perspectives on forest, wildlife conservation and, crucially, forest restoration. Such a fundamental shift in thinking will not be possible without recognizing and embracing the immense reservoir of forest knowledge possessed by communities. "I don't want to use the term 'indigenous' knowledge," says forester Komol Praekthong, "because that makes it sound like something in the past. We are talking about knowledge that is like a science. It is evolving all the time."[7]

It is of course possible that deforestation will continue even after foresters and forest communities have made peace. But without resolution of this century-old conflict, loss of forest is almost certain to continue until there is little left to protect. A precondition for this peace is a brutally honest reexamination of state forestry's place in the sad story of deforestation in Thailand. There is no shame in such an exercise; nor does it suggest that others are blameless. But out of such a review, creative, genuine solutions to intractable conflicts are likely to emerge. Thai forestry history suggests that there are no quick-fix models out there that can be imported and applied. It will certainly not be enough simply to copy the "new forestry" that is practiced in the Pacific Northwest and Northern Europe. A new Thai forestry will have to be invented, one based on the cultural, biological, and political conditions that prevail in Thailand (with intelligent adaptations from others' experiences) and on Thai foresters' willingness to reinvent themselves.

NOTES

INTRODUCTION

1. Personal communication with Somsak Sukwong, founding director of Regional Community Forestry Training Center for Asia and the Pacific (RECOFTC), 1991.
2. William A. Smalley, *Linguistic Diversity and National Unity: Language Ecology in Thailand* (Chicago: The University of Chicago Press, 1994), 1.
3. Ibid., 14. Major regional languages like Kammueang (north), Lao (northeast) and Phaktai (south) are spoken natively by more than 20 million people, and are as distant in mutual intelligibility from Standard Thai as Bengali, Gujarati, and Marathi are from Hindi in multi-lingual India. What William Smalley calls "marginal regional languages"—Northern Khmer, Pattani Malay, and two Karen languages, Sgaw and Pwo—are spoken by 3,400,000 people and are as different from Standard Thai as is English. Enclave languages make up the rest, spoken in small pockets by as few as several dozen people.
4. Kamala Tiyavanich, *Forest Recollections: Wandering Monks in Twentieth-Century Thailand* (Chiang Mai: Silkworm Books, 1997), 249.
5. A breakdown of key dates up to 1998 is provided in Verena Brenner, et al., "Thailand's Community Forest Bill: U-turn or Roundabout in Forest Policy?," SEFUT Working Paper No. 3, rev. ed. (University of Freiburg, January 1999), 15–21.
6. Borwornsak Uwanno and Wayne D. Burns, "The Thai Constitution of 1997: Sources and Process," *University of British Columbia Law Review* 32, no. 2 (1998): 227–33.
7. Withaya Aphorn, "Kreuakhai khong pa chumchon nai prathet thai pi 2004, Krongkarn Sitthichumchonseuksa," *Trang* (April 2004).
8. Interview with Somying Soontornwong, January 2008.
9. This occurred three months after the 1997 Constitution was abrogated by the coup-makers and replaced by a new constitution in August 2007. Though the new constitution reversed many of the achievements of the previous document in the areas of civil rights and state accountability, it was more progressive on the question of community rights and managing natural resources.
10. D. Feeney, "Agricultural Expansion and Forest Depletion in Thailand, 1900-1975," in *World Deforestation in the Twentieth Century*, eds. J. Richards and R. Tucker (Durham and London: Duke University Press, 1988), 112–287; Royal Forestry Department, *Forest Statistics of Thailand 1999* (Bangkok: Royal Forestry Department, 1999): 4.

11 The RFD now has jurisdiction over about 60 million rai (92,000 cu sq km). Interview with Upai Wayupat, deputy director general of the Royal Forestry Department, January 2008 (with thanks to Nantiya Tangwisuttijit).

12 D. Bourke-Borrowes, "General Thoughts and Observations on Forestry in Siam," *Indian Forester* 54, no. 3 (1928): 142.

13 Interview with Niels Elers Koch, director of the Danish Centre for Forest Landscape Planning, December 1999.

14 See, e.g., Mac Chapin, "A Challenge to Conservationists," *World Watch Magazine* (November–December 2004); Gill Shepherd, et al., "Rights, Tenure, Governance and a More Pro-Poor Vision for Conservation: What Should We Be Aiming At?" (background paper for the conference *Towards a New Global Forest Agenda*, Stockholm, October 29, 2007).

PART I. WATERSHEDS OF THAI FORESTRY HISTORY

1 Interview with Veerawat Dheeraprasart of the Foundation for Ecological Recovery, January 2008.

2 Belinda Stewart Cox, "Thailand's Nam Choan Dam—A Disaster in the Making?" *The Ecologist* 17, no. 6 (1987): 212–19.

3 Team Consulting Co. Ltd., *Environmental and Ecological Impacts* (Bangkok, 1980).

4 Prinya Nutalai, "Earthquakes and the Nam Choan Dam," *The Ecologist* 17, no. 6 (1987): 223–25. According to Prinya, a geologist, earthquakes were also reported in the area of the proposed dam site in 1785, 1912, 1930, and 1959.

5 Team Consulting, *Environmental and Ecological Impacts*.

6 Interview with Choompol Ngamponsi, Kasetsart University forestry professor, March 1988.

7 Ann Danaiya Usher, "To measure a forest . . . 1987," *Nam Choan Inquiry: The Environmental Dilemma of the Decade*, a supplement of *The Nation*, Bangkok, April 18, 1988.

8 Interview with Subin Panyamag from the public relations department of the Electricity Generating Authority of Thailand (EGAT), March 1988.

9 Interviews with members of the ground surveyors' team consisting of three Third and Fourth Year Kasetsart University Forestry Faculty students, February 1988. All spoke on condition of anonymity fearing reprisal from professors during exams.

10 Interviews with Karen guides in Thung Yai Naresuan Wildlife Sanctuary, February 1988.

11 *The Nation*, February 25, 1988.

12 Interview with Pairoj Suwannakorn, director of Royal Forestry Department's Wildlife Conservation Division, June 1993.

13 Interview with Bhichai Rattakul, August 1992.
14 "MP says seven ministers are involved in logging," *Bangkok Post*, December 14, 1988.
15 "Students say they will sue if loggers are not punished," *Bangkok Post*, December 15, 1988.
16 "Kukrit calls for withdrawal of all concessions," *Bangkok Post*, December 6, 1988. Kukrit was right on the first count, but wrong on the second. Chatichai Choonhavan's government lasted just two and a half years before being ousted in the coup d'etat of February 23, 1991.
17 Tunya Sukpanich, "Villagers protest against logging firms in North," *Bangkok Post*, April 25, 1988; "Wanted: A strict policy on logging concessions," *Bangkok Post*, December 8, 1988.
18 This legal advisory body's role was to interpret contentious points of law, but it had no authority to impose its rulings.
19 A report on the causes and implications of the floods of 1988 concluded that three- to five-year-old rubber trees planted indiscriminately in response to government subsidies and a ready foreign market were "the most dramatic land use change ... [which was] the single most important human activity that contributed to the disaster." See National Economic and Social Development Board and the United States Agency for International Development, *Safeguarding the Future: Restoration and Sustainable Development in the South of Thailand* (Bangkok, 1989).
20 "Valley of death," *Bangkok Post*, November 26, 1988.
21 Thiwa Saphakij speaking at a conference organized for pulp and paper companies by the Thai Forestry Sector Master Plan team, Bangkok, February 1992.
22 Interview with Uthit Kut-In, dean of forestry, Kasetsart University, February 1999.
23 Bichai Rattakul interview.
24 Visit to Huai Kaeo, March 1992.
25 Interview with Kitti Damnerncharnvanich, owner of Suan Kitti Reforestation Company, February 1990.
26 Shell Thailand intended to obtain permission to rent out 2,000 hectares of government forest land to establish eucalyptus plantations. This would have required cabinet approval.
27 Kitti Damnerncharnvanich interview.
28 Ann Danaiya Usher, "What price the eucalyptus tree?," *The Nation*, February 22, 1990.
29 Ann Danaiya Usher, "The dilemma of the 1990s: Eucalyptus vs. land rights," *The Nation*, May 5, 1989.

30 Boonwong Thaiutsa, "Mai eucalyptus" [The Eucalyptus Tree] (unpublished paper, 1990), 3.
31 Interview with Reungchai Pao-sujja, head of Private Reforestation Promotion Office, Royal Forestry Department, March 1992.
32 The *Khor Chor Kor* was officially revoked on June 23, 1992.
33 Thai NGO Committee on Forest and Land Problems in Northeast Thailand, "The Programme for Agricultural Land Distribution for Poor People Living in Degraded Forest Reserves (*Khor Chor Kor*): Questions and Answers" (unpublished paper, 1992).
34 Interview with Sisuwan Kuankachorn, September 1992. Sisuwan's organization, Project for Ecological Recovery, was among the NGOs that lobbied government ministries and the cabinet on behalf of villagers in the crucial weeks before the cancellation of *Khor Chor Kor*.
35 Sanitsuda Ekachai, "Torn from the land: how tree-planting uproots whole villages," *Bangkok Post*, January 23, 1992.
36 Ann Danaiya Usher, "In the hours after the arrest," *The Nation*, September 8, 1991. See also, e.g., Chainarong Setthachuea and Malee Traisawasdichai, "Evicted," *The Nation*, January 22, 1992; Sanitsuda Ekachai, "Man and the forest," *Bangkok Post*, January 24, 1992; Malee Traisawasdichai, "To live and die for a home," *The Nation*, June 17, 1992.
37 Ann Danaiya Usher, "In the hours after the arrest," *The Nation*, September 8, 1991.
38 Sri Plingkatok, a fifty-two-year-old matriarch, quoted in Sanitsuda Ekachai, "Torn from the land: how tree-planting uproots whole villages" *Bangkok Post*, January 23, 1992.
39 Sanitsuda Ekachai, "More than a question of land rights," *Bangkok Post*, June 23, 1992.
40 Interview with forester Krishna Brikshavana, October 1999.
41 This section is adapted from two articles I wrote at the time of Seub's death: "Last farewell to a dedicated conservationist," *The Nation*, September 6, 1990, and "Visions of Seub Nakhasathien," *The Nation*, September 10, 1990.
42 The song's title is "Seub Nakhasathien," written by Yuenyong Ophakul (more commonly known as At Carabao).

PART 2. SCIENTIFIC FORESTRY ENTERS SIAM

1 Letter from King Chulalongkorn to Chao Muang Inthanon, quoted in James A. Ramsay, "The Development of a Bureaucratic Polity: The Case of Northern Thailand" (Ph.D. diss., Cornell University, 1971), 150.
2 Thongchai Winichakul, *Siam Mapped: A History of the Geo-body of a Nation*, (Chiang Mai: Silkworm Books, 1994), 13.

3 Ian Brown, *The Élite and the Economy in Siam, c. 1890–1920* (Singapore: Oxford University Press, 1988).

4 Raymond Bryant, "From Laissez-faire to Scientific Forestry: Forest Management in Early Colonial Burma, 1826–1885," *Forest and Conservation History* 34, no. 4 (October 1994): 160–70.

5 E. P. Stebbbing, *The Forests of India*, vol. 1 (London: John Lane The Bodley Head Ltd., 1921), 37.

6 Herbert Slade, "Report of the Royal Forestry Department for 119 (1901)," National Archives of Thailand, MR 5 M/41/1: 51.

7 Henry E. Lowood, "The Calculating Forester: Quantification, Cameral Science, and the Emergence of Scientific Forestry Management in Germany," in *The Quantifying Spirit of the 18th Century*, eds. Tore Frängsmyr, J. L. Heilbron, and Robin E. Rider (Berkeley: University of California Press, 1990), 341.

8 Ibid., 334.

9 Niels Elers Koch and James J. Kennedy, "Multiple-Use Forestry for Social Values," *Ambio* 20, no. 7 (1991): 331.

10 Hamilton King, "Teak Industry of Siam," *USCR* 66 (July 1901): 305–10. (With thanks to Suthee Prasartset.)

11 H. Colyear Dawkins and Michael S. Philip, *Tropical Moist Forest Silviculture and Management: A History of Success and Failure* (Oxon: CAB International, Wallingford, 1998), 14–19.

12 Stebbing, *Forests of India,* 34–37.

13 Similar courses would soon follow in Saxony at Tharandt, southwest of Dresden, in Prussia at Eberswalde, later to become a part of the University of Berlin, and at the Hannoversch-Münden Academy, which became part of Göttingen University in 1926. See Kurt Mantel, "History of the International Science of Forestry with Special Consideration of Central Europe: Literature, Training, and Research from the Earliest Beginnings to the Nineteenth Century," in *International Review of Forestry Research*, vol. 1, eds. John A. Romberger and Peitsa Mikola (New York: Academic Press, 1964), 19; Robert Winters, *The Forests and Man* (New York: Vantage Press, 1974), 275; interview with Siegfried Lewark, German forestry professor, April 1999.

14 Franz Heske, *German Forestry* (New Haven: Yale University Press, 1938), 81.

15 Lowood, "The Calculating Forester," 333–38.

16 Ibid., 329.

17 James C. Scott, *Seeing Like a State: How Certain Schemes to Improve the Human Condition Have Failed* (New Haven and London: Yale University Press, 1998), 11–20.

18 Lowood, "The Calculating Forester," 318.

19. Hermann Graf Hatzfeldt. "Forestry in Germany: Old and New" (unpublished paper, 1996).
20. Siegfried Lewark interview.
21. This general emphasis in German university forestry training remained until the 1990s. See Siegfried Lewark, "The Radical Revision of the Forestry Curriculum at the Freiburg Forestry Faculty" (unpublished paper delivered at FAO Advisory Committee on Forestry Education [ACFE] Meeting, Chile, 1996.)
22. Stebbbing, *Forests of India*, 367–68.
23. Gifford Pinchot, *Breaking New Ground* (New York: Harcourt, Brace, & Co., 1974), 19.
24. Dietrich Brandis, *Notes on Forest Management in Germany* (Calcutta: Office of the Superintendent of Government Printing, India, 1866), Office of India Records, 1866, V/27/560/136: 1.
25. Stebbbing, *Forests of India*, 372.
26. W. R. Fisher, "Review on the New Edition of Volume IV of Dr. Schlich's *Manual of Forestry: Forest Protection*." *Indian Forester* 33, no. 8 (1907): 351.
27. Dawkins and Philip, *Tropical Moist Forest*, 55.
28. W. R. Fisher, *Schlich's Manual of Forestry*, vol. 4, *Forest Protection*, trans. and adapted from "Der Forstschutz" by Dr. Richard Hess (London: Agnew & Co. Ltd.), 1907.
29. E. E. Fernandez, *Rough Draft of a Manual of Indian Silviculture* (Dehra Dun: Indian Forest Service, 1891), 4–5.
30. R. S. Troup, "A Note on Some European Sylvicultural Systems, with Suggestions for Improvements in Indian Forest Management" (Calcutta: Office of the Superintendent of Government Printing, India, 1916), Office of India Records, 1916, V/27/560/35: 4.
31. In 1886 Bernard Fernow was the only professional forester in the United States and became chief of the division of forestry in the Department of Agriculture. He founded forestry education in the United States at Cornell University and in Canada at the University of Toronto. Soon thereafter an Austrian-trained forester, John A. Warder, established the country's first forestry organization, the American Forestry Association. See Craig W. Allin, *The Politics of Wilderness Preservation* (Westport, CT: Greenwood Press, 1982), 29
32. Bernard Fernow, *The Economics of Forestry* (New York: Thomas Y. Crowell, 1902), 229.
33. Marion Clawson and Roger Sedjo, "History of Sustained-Yield Concept and Its Application to Developing Countries" in *History of Sustained Yield Forestry: A Symposium*, ed. Harold K. Steen (Durham: Forest History Society, 1984), 7. See also Nancy Langston's excellent *Forest Dreams, Forest Nightmares* (Seattle: University of Washington Press, 1995), 104–13.

34 James Kennedy, Michael P. Dombeck, and Niels E. Koch, "Values, Beliefs and Management of Public Forest in the Western World at the Close of the Twentieth Century," in *New Requirements for University Education in Forestry: Proceedings of a Workshop Held in Wageningen, The Netherlands, 30 July–2 August, 1997*, eds. P. Schmidt, J. Huss, S. Lewark, D. Pettenella, and O. N. Sastamoinen, (Korbeek-Dijle: Drukkerij De Weide 1998), 17.

35 R. Ribbentrop, *Forests in British India* (Calcutta: Office of the Superintendent of Government Printing, India, 1900), Office of India Records, 1900, V/27/560/8: 106.

36 Wilhelm Schlich, *Schlich's Manual of Forestry*, vol. 1, *Forest Policy in the British Empire*, 4th ed. (London: Bradbury, Agnew & Co. Ltd., 1922), quoted in Office of India Records, 1922, V/27/561/2.

37 *Report on Forest Administration in Burma for 1922–23* (Glossary) (Rangoon: Office of the Superintendent of Government Printing, Burma, India Office, 1923), Office of India Records, 1923, V/24/1397.

38 Stebbing, *The Forests of India*, 36.

39 Ramachandra Guha, "Early Environmentalists in India: Some Historical Precursors" (Foundation Day lecture, Society for Promotion of Wastelands Development, New Delhi, June 15, 1993), 8.

40 Pinchot, *Breaking New Ground*.

41 Dietrich Brandis, *Suggestions Regarding Forest Administration in British Burma, January 25, 1876* (Calcutta: Office of the Superintendent of Government Printing, India, 1876), Office of India Records, 1876, V/27/560/72: 3–4.

42 Ramachandra Guha, "An Early Environment Debate: The Making of the 1878 Indian Forest Act," *Indian Economic and Social History Review* 27, no. 1 (January–March 1990): 65–84.

43 Dietrich Brandis, "The Progress of Forestry in India," *Indian Forester* 10, no. 11 (1884): 509.

44 H. C. Walker, "Scientific Forestry," *Indian Forester* 33, no. 10 (1907): 452–56.

45 This fourth volume was, in fact, written by a Dr. Richard Hess, Professor of Forestry at the University of Giessen, reportedly one of Germany's best-known foresters. See Fisher, *Schlich's Manual of Forestry*, vol. 4.

46 W. R. Fisher, "Review of Dr. Schlich's *Manual of Forestry*," 351.

47 H. C. Walker, "Fire Protection in Burma." *Indian Forester* 34, no. 6 (1908): 339–49.

48 G. R. Long, "Displacement of Taungya Plantations by Improvement Fellings." *Indian Forester* 33, no. 6 (1907): 281–82.

49 Ibid.

50 Herbert Slade, "Too Much Fire-Protection in Burma," *Indian Forester* 22, no. 5 (1896): 172–76. All quotes by Slade on pages 49–50 are attributable to this article. (With special thanks to Raymond Bryant for bringing this article to my attention.)

51 F. Beadon Bryant, "Fire Conservancy in Burma," *Indian Forester* 33, no. 12 (1907): 538.

52 Slade, "Too Much Fire-Protection."

53 Ibid.

54 Slade, "Report of the Royal Forestry Department."

55 Reginald Le May, *An Asian Arcady: The Land and Peoples of Northern Siam* (Cambridge: W. Heffer & Sons Ltd., 1926), 54. Chaiyan Rajchagool finds the use of the metaphor "silent revolution" inappropriate given the two instances of violent resistance in Chiang Mai itself and the surrounding areas. He does, however, subscribe to the theory that administrative reform in Lanna was the result of cooperation between Bangkok and the British for the benefit of the teak industry. See Chaiyan Rajchagool, *The Rise and Fall of the Thai Absolute Monarchy: Foundations of the Modern Thai State from Feudalism to Peripheral Capitalism* (Bangkok: White Lotus, 1994), 95.

56 Le May, *Asian Arcady*, 62.

57 W. F. L. Tottenham, "The Formation of the Forest Department in Siam," *Indian Forester* 31, no. 8 (1905): 448.

58 David F. Holm, "A History of the Teak Industry in Thailand" (unpublished paper, Yale University, May 1969); Brown, *The Élite and the Economy in Siam*, 118.

59 Brown, *The Élite and the Economy in Siam*, 94.

60 Royal Forestry Department, *Prawat krom pa mai, 2439–2514* (Bangkok, 1971), 10.

61 A. C. Pointon, *Bombay-Burmah Trading Corporation Limited 1863–1963* (Southampton: The Millbrook Press Limited, 1964), 34.

62 Le May, *Asian Arcady*, 61.

63 Secret 1902 report by the British Consul in Chiang Mai, quoted in Ramsay, "Development of a Bureaucratic Polity," 121.

64 Banasopit Mekvichai, "The Teak Industry in North Thailand: The Role of a Natural Resource-Based Export Economy in Regional Development" (Ph.D. diss., Cornell University, 1988), 206.

65 Ramsay, "Development of a Bureaucratic Polity," 120.

66 Herbert Slade, "Rai-ngan khong Mr. H. Slade rueang kan pa mai khong prathet sayam" (repr., Bangkok: Information Department of the Office of Forestry Secretariat, 1986), 41.

67 Ibid., 16.

68 The first treaty with a foreign power relating specifically to the north of Siam was signed in Calcutta in 1874, two years after King Chulalongkorn's first visit to India, marking a first attempt to exercise direct control over the affairs of the *chao*. It was replaced by the treaty between Great Britain and Siam in 1883, which included the same rules for girdling and logging, as well as the "wistful note" about keeping the northern princes in line. The treaty also provided for a British Consul to reside in Chiang Mai. See Le May, *Asian Arcady*, 55.

69 Ibid., 61.

70 Interview with Plodprasob Suraswadi, director general of the Royal Forestry Department, October 1999.

71 In contrast, both natural and cultivated teak in Java tends to dominate the forest, making natural teak forest appear virtually the same as monoculture plantation forest. See Nancy Peluso, *Rich Forests, Poor People: Resource Control and Resistance in Java* (Berkeley: University of California Press, 1994).

72 Stebbbing, *Forests of India*, 377.

73 D. Bourke-Borrowes, "The Teak Industry of Siam," *Technical and Scientific Supplement to the Record* (Bangkok: Ministry of Commerce and Communications, 1927), 155.

74 Girdling is defined by foresters as "a method of killing teak and other heartwood trees by ringing the sapwood." See the glossary in *Report on Forest Administration in Burma for 1922–23* (Rangoon, Office of the Superintendent of Government Printing, Burma, India Office, 1923), Office of India Records, 1923, V/24/1397.

75 Banasopit Mekvichai, "Teak Industry," 130.

76 Ramsay, "Development of a Bureaucratic Polity," 138; Suree Phumiphamorn, "Maisak mueang sayam: Chut kamnot krom pa mai," in *Nitiyasarn Silapawattanatham*, (mimeograph, n.d.), 137.

77 Personal communication with Niels K. Dyrlund, Danish ambassador to Thailand, November 1999.

78 Brown, *The Élite and the Economy in Siam*.

79 *British Burmah Revenue and Agriculture Proceedings: Forests, January to June 1896*, Part II-B, file no. 2A 7 (January 1896): viii. Slade remained in Siam until 1901, then returned to Burma where he died in 1905.

80 Slade, "Rai-ngan," 27.

81 Ibid., 11.

82 Ibid., 41.

83 Bourke-Borrowes, "Teak Industry of Siam," 157–58; Slade, "Report of the Royal Forestry Department," 21.

84 Pointon, *Bombay-Burmah Trading Corp.*, 37.

85 Bourke-Borrowes, "Teak Industry of Siam," 153.

86 Chamaichome Sunthornsawasdi, *Historical Study of Forestry in Northern Thailand Between 1896 and 1932* (Bangkok, 1978), 128

87 LeMay, *An Asian Arcady*, 62.

88 A Thai forestry training school would not be established until 1935. It was located in Phrae Province at the former headquarters of the Danish East Asiatic Company.

89 Slade, "Report of the Royal Forestry Department," 4.

90 Plodprasob Suraswadi interview.

91 Royal Forestry Department, *Prawat krom pa mai 2439–2500* (Bangkok, 1958), 32.

92 Tottenham, "Forest Department in Siam," 448.

93 Ibid., 447.

94 Slade, "Report of the Royal Forestry Department," 40.

95 Ibid., 16.

96 Herbert Slade, "The Maihongson Forests in Siam," *Indian Forester* 27, no. 5 (1901): 477, quoting a report by Special Deputy in Chiang Mai J. G. F. Marshall.

97 Slade, "Report of the Royal Forestry Department," 16.

98 Ibid., 20.

99 Ibid., 7.

100 Ramsay, "Development of a Bureaucratic Polity," 261–64.

101 Royal Forestry Department, *Forestry Statistics of Thailand 1998* (Bangkok: Data Center, Information Office, Royal Forestry Department, 1998), 11.

102 Chamaichome Sunthornsawasdi, *Historical Study of Forestry*, 39–40.

103 Holm, "History of the Teak Industry," 46.

104 Banasopit Mekvichai, "Teak Industry," 228.

105 Boonsong Lekagul, *Hak pa mai yang yu yang yuen yong* (1959), 135.

106 Banasopit Mekvichai, "Teak Industry," 225.

107 Ibid., 230.

108 Twenty-eight percent is the official figure, but the reality was probably much less.

109 Slade, "Report of the Royal Forestry Department," 23.

110 Virginia Thompson, *Thailand: The New Siam* (New York: The MacMillan Company, 1941), 833.

111 Ramsay, "The Development of a Bureaucratic Polity," 125–26.

112 Le May, *An Asian Arcady*, 58.

113 Suree Phumiphamorn, "Maisak mueang sayam," 136.

114 Ramsay, "The Development of a Bureaucratic Polity," 126.

115 "Statement of the Case of the Estate of the Late Dr. Marion Alonzo Cheek Against the Siamese Government," National Archives Thailand, file no. MR 5 M/41/2: 1–2.

116 Brown, *The Élite and the Economy in Siam*.

117 Thompson, *Thailand*, 835.

118 Ramsay, "The Development of a Bureaucratic Polity," 126.

119 Pointon, *Bombay-Burmah Trading Corp.*, 35.

120 Thompson, *Thailand*, 835.

121 Interview with Ole Lange, Danish economic historian, December 1999.

122 Pointon, *Bombay-Burmah Trading Corp.*, 35.

123 Ibid., 37.

124 Le May, *An Asian Arcady*, 58.

125 East Asiatic (Thailand) Public Co. Ltd., *A Tale of Two Kingdoms* (Bangkok, 1995), 246–62.

126 Thompson, *Thailand*, 834–35.

127 Pointon, *Bombay-Burmah Trading Corp.*, 34; Le May, *An Asian Arcady*, 58.

128 Le May, *An Asian Arcady*, 58.

129 Interview with Jorni Odechao, Karen community elder, October 1999.

130 Bourke-Borrowes, "General Thoughts," 142.

131 Banasopit Mekvichai, "Teak Industry," 252.

132 Chaiyan Rajchagool, *Rise and Fall of the Thai Absolute Monarchy*, 94–95.

133 D. O. Witt, "The Use and Abuse of Forest Work in Siam," *Indian Forester* 30, no. 7 (1904): 299–303, quoted in Oliver Pye, *Khor Jor Kor: Forest Politics in Thailand* (Bangkok: White Lotus, 2005), 30.

134 Nancy Lee Peluso and Peter Vandergeest, "Genealogies of Forest Law and Customary Rights in Indonesia, Malaysia and Thailand" (draft paper, 1999), 55.

135 Ibid., 49.

136 Jorni Odechao interview.

137 Ibid.

138 Author's rough English translation based on Jorni's translation into Thai.

139 *Progress Report of Forest Administration in British Burma, 1865–66*, quoted in Raymond Bryant, "From Laissez-faire to Scientific Forestry: Forest Management in Early Colonial Burma, 1826–1885," *Forest and Conservation History* 34, no. 4 (October 1994): 166.

140 Ibid., 168.

141 Dietrich Brandis, *Indian Forestry* (1897): 77, quoted in K. Sivaramakrishnan, "The Politics of Fire and Forest Regeneration in Colonial Bengal," Special issue: South Asia, *Environment and History* 2, no. 2 (June 1996): 156.

142 Ramachandra Guha, *The Unquiet Woods: Ecological Change and Peasant Resistance in the Himalaya* (Delhi: Oxford University Press, 1989), 110.

143 J. C. Nelson, "Forest Settlement Report of the Garhwal District" (Lucknow, 1916), 10–11, quoted in Guha, *The Unquiet Woods*, 106.

144 Peluso, *Rich Forests, Poor People*.

145 Ibid., 69–72. Here, Peluso refers specifically to the most visible of the rural protesters in the teak forests of Java at the turn of the nineteenth century, the Samin Movement.

146 Ibid., 67.

147 Banasopit Mekvichai, "Teak Industry," 263–66.

148 Interview with Nuan Lachai, community leader in Samoeng, April 1988.

149 Colonial foresters divided forest timber trees into two types: teak and non-teak hardwoods. *Mai krayaloei* (hardwoods other than teak) were referred to in English as "otherwood" trees. See Bourke-Borrowes, "The Teak Industry of Siam."

150 Witoon Permpongsacharoen, "Tropical Forest Movements: Some Lessons from Thailand," in *Volume on Tropical Forests*, ed. Heffa Schücking (Sassenberg: Arbeit Gemeinschaft Regenwald 1991.

151 Villagers' testimonies at a meeting organized by the Union of Civil Liberties and the Project for Ecological Recovery, April 1988.

152 Interview with Winai Subrungruang, former deputy managing director, Forestry Industry Organization (FIO) October 1999.

153 Montree Janthawong, et al., "Rai-ngan phon kan wichai rueang khwamlaklai thang chiwaphap lae rabop niwet nai khet pa chumchon phak nuea ton bon" [Report of Research Results on Biodiversity and Ecological Systems in Community Forests in the Upper North] (unpublished study, n.d.), 115–37.

154 Udom Charoenniyomphrai et al., eds. *Research Report on Indigenous Knowledge, Customary Use of Natural Resources and Sustainable Biodiversity Management: Case Study of Hmong and Karen Communities in Thailand* (Chiang Mai: Inter Mountain Peoples Education and Cultures in Thailand Association [IMPECT], June 2006), 73.

155 "Rotational agriculture is just a nice way of saying it," said Chalermsak Vanichsombat, the director general of the National Parks Department. "Really it is slash and burn. They slash and then burn one area of forest and then move on to another. You can't just say that this is a 'way of life' or that it is 'traditional'. Fifty years ago it was okay because there was lots of forest and very few people. But society is changing and they have to change too." Interview with Chalermsak Vanichsombat, director general of the National Parks Department, January 2008.

156 Bourke-Borrowes, "General Thoughts," 157–58.

157 Ibid.

158 W. F. L. Tottenham, "Forestry," in *The Kingdom of Siam 1904*, ed. A. Cecil Carter (Bangkok: The Siam Society, 1988), 177.

159 Tottenham, "Forest Department in Siam," 448.

160 D. Bourke-Borrowes became advisor to the Royal Forestry Department in 1924, when Phraya Daruphanphitak became the first Thai to take charge of the department, replacing Siam's third British conservator of forests, W. F. Lloyd.

161 Bourke-Borrowes, "General Thoughts," 157.

162 Banasopit Mekvichai, "Teak Industry," 182, 186, 191.

163 Ibid., 167.

164 Interview with Prasert Prachit of the Forestry Industry Organization (FIO), October 1999. The minimum exploitable girth for hardwoods other than teak was 80 cm.

165 Banasopit Mekvichai, "Teak Industry," 185–86.

166 Y. S. Rao, "Thailand's Teak Forests Today," *Asian Timber* (January–February 1982): 42–45.

167 F. Loetsch, *Report to the RTG on Forest Inventory of the Northern Teak Bearing Provinces* (Rome: Food and Agriculture Organization, 1958).

168 Constance M. Wilson, *Thailand: A Handbook of Historical Statistics* (Boston: G. K. Hall & Co., 1983), 137–38.

169 Royal Forestry Department, *Forestry Statistics of Thailand 1985* (Bangkok: Planning Division, Royal Forestry Department, 1985): 26–27, 66–77.

170 Apichart Kaosa-ard, Verapong Suangtho, and Erik Kjær, "Teak (*Tectona grandis* Linn. f.)," in *ASEAN: A Survey Report* (Bangkok: ASEAN/Canada Forest Seed Centre, October 1986), 19.

171 This table is based on a compilation of tables presented in Banasopit Mekvichai, "Teak Industry," 182, 186, 191. She derives figures for 1896–1926 from Bourke-Borrowes, "The Teak Industry;" for 1935 from Central Service of Statistics, *Statistical Yearbook, Siam, No. 19*, 1935/36–1936/37 and *The Bangkok Times Press*, Bangkok, 1939; for 1940 from Central Service of Statistics, *Statistical Yearbook, Siam, No. 21*, 1939/40–44 and University of Moral and Political Sciences Press, Bangkok, 1950; and lastly for 1950–1978 from Forestry Industry Organization, *Thiraluk khroprop 30 pi*, Bangkok.

172 Plodprasob Suraswadi interview.

173 Interview with Apichart Kaosa-ard, Thailand's teak expert, October 1999.

174 Ibid.

175 Veerawat Dheeraprasart interview.

176 Dawkins and Philip, *Tropical Moist Forest*, 45.

177 Apichart Kaosa-ard interview.

178 Boonsong Lekagul, *Hak pa mai*.

179 Banasopit Mekvichai, "Teak Industry," 143.
180 Ibid., 143.
181 Ibid., 140.
182 Winai Subrungruang interview.
183 Royal Forestry Department, *Forestry Statistics of Thailand 1998* (Bangkok: Data Center, Information Office, Royal Forestry Department, 1998), 15.
184 Ibid.
185 Banasopit Mekvichai, "Teak Industry," 145.
186 "Forest officials' death part of a bloody balance sheet," *The Nation*, April 15, 1993.
187 Anat Arbhabhirama, et al., *Thailand Natural Resources Profile* (Bangkok: Thailand Development Research Institute, 1987), 75.
188 Royal Forestry Department, *Forest Statistics of Thailand 1998* (Bangkok: Data Center, Information Office, Royal Forestry Department, 1998), 13
189 Banasopit Mekvichai, "Teak Industry," 125.
190 Ibid.
191 Bourke-Borrowes, "General Thoughts," 156.
192 Ibid., 147–48.
193 Dusit Banijbatana, "Forest Policy in Northern Thailand," in *Farmers in the Forest: Economic Development and Marginal Agriculture in Northern Thailand*, eds. Peter Kunstadter, E. C. Chapman, and Sanga Sabhasri (Honolulu: East-West Center, University Press of Hawai'i, 1978), 58.
194 Apichart Kaosa-ard, et al., "Teak." FIO's teak plantations now cover approximately 924 sq km.
195 B. Boontawee, "Thailand Country Report on State of Forestry" (report presented at the 18th Session of the Asia-Pacific Forestry Commission, FAO, Noosaville, Queensland, Australia, May 15–19, 2000), 13.
196 Bourke-Borrowes, "General Thoughts," 143.
197 Ibid., 145.
198 Ribbentrop, *Forests in British India*, 166–67.
199 Troup, "Note on Some European Sylvicultural Systems."
200 Ibid.
201 R. S. Troup, *The Silviculture of Indian Trees*, vol. 1 (Oxford: Clarendon Press, 1921) vi.
202 Banasopit Mekvichai, "Teak Industry," 243.
203 Winai Subrungruang interview.
204 Banasopit Mekvichai, "Teak Industry," 244.

205 Dusit, "Forest Policy," 58.

206 Winai Subrungruang interview.

207 Long, "Displacement of Taungya Plantations," 281–82.

208 Herbert Slade, "Report of the Royal Forestry Department," 8. While this statement reflects the common attitude of foresters, it contradicts the impressions Slade gives in his 1896 article in *Indian Forester*, where he suggests that teak forests had co-existed with human settlements and annual fires since the "remote ages."

209 Ibid., 7.

210 Ibid., 61.

211 In interviews, both Tongroj Onchan and Krishna Brikshavana claimed to have been the ones to come up with the 10 million figure.

212 Ministry of Agriculture, *Classification of Uses of Forest Land in National Forest Reserves* (Bangkok: National Forest Resource Management Unit of the National Forest Resource Lang Management Division, Royal Forestry Department, 1992), 32.

213 Krishna Brikshavana interview.

214 Ibid.

215 Kultida Samabuddhi, "House blasted for tardy action on forests bills," *Bangkok Post*, December 22, 2004.

PART 3. THE LOGICAL CONCLUSION: FACTORY FORESTS

1 Ribbentrop, *Forests in British India*, 167.

2 Dawkins and Philip, *Tropical Moist Forest*, 43.

3 Julian Evans, *Plantation Forestry in the Tropics: Tree Planting for Industrial, Social, Environmental and Agroforestry Purposes* (Oxford: Clarendon Press, 1992), 340–41.

4 Apichart Kaosa-ard interview.

5 The Forest Herbarium, "Vegetation Types of Thailand" (http://www.dnp.go.th/Botany/Flora/Forest_type.htm) reports 2,500 tree species, while Harvard University botanist, Peter Ashton, estimates the number could range from 3,000 to 3,500. Personal communication with Peter Ashton.

6 It is surely no coincidence that these same three types of trees account for 85 percent of all tree plantations in the tropics. Evans, *Plantation Forestry*, 36.

7 The Food and Agriculture Organization (FAO), UN Environment Program (UNEP), and UN Development Program (UNDP) as well as the bilateral aid agencies of Japan, Canada, Australia, Finland, and Sweden have all provided aid that has supported the planting of teak, pine, and/or eucalyptus trees.

8 East Asiatic, *Two Kingdoms*, 30.

9 See Leif Svalesen, *The Slave Ship Fredensborg* (Bloomington: Indiana University Press, 2000).
10 Ole Lange interview.
11 East Asiatic, *Two Kingdoms*, 32.
12 Ibid., 34.
13 Interview with Jens Granhof, October 1999.
14 East Asiatic, *Two Kingdoms*, 36.
15 Ibid., 37.
16 Ibid., 39.
17 Ole Lange interview.
18 Niels K. Dyrlund personal communication.
19 The name Andersen and Co. Ltd. was changed to East Asiatic Company in 1897. While Siam was the company's "cradle" and teak its initial focus, after Andersen sold the Oriental to Louis Leonowens, he used the profits to buy two sailing ships. The company's business would eventually expand to include a steamship line from Denmark to Siam via Singapore, and various other industrial and trading activities around the world. The best-known myth about Andersen is that he used the Oriental as a brothel. According to Ole Lange, Danish economic historian and author of the two-volume study of H. N. Andersen, this never had any basis in fact. As Lange's research shows, during the years that Andersen owned the hotel, it was the permanent residence of a group of Danish sailors living in Bangkok and their Thai lovers. But during the First World War, by which time Andersen had become the king of Denmark's main foreign policy adviser and Denmark's "unofficial foreign minister," the British government also wondered about the truth of rumors that the Dane had in the 1880s been a brothel owner. The British Consul in Bangkok investigated the matter and came to the conclusion that the Oriental deteriorated into a brothel under the later ownership of Louis Leonowens. Ole Lange interview; Ole Lange, *Den hvide elefanten: H. N. Andersens eventyr og ØK 1852–1914*, Gyldendal, Copenhagen, 1986; Ole Lange, *Jorden er ikke større . . . H. N. Andersen, ØK og storpolitikken 1914–37*, Gyldendal, Copenhagen, 1988.
20 Lange, *Jorden er ikke større*, 35.
21 East Asiatic, *Two Kingdoms*, 41.
22 Ole Lange interview.
23 Prince Axel was a grandson of Denmark's King Christian IX (AD 1818–1906). See East Asiatic, *Two Kingdoms*, 190.
24 Ibid., 246.
25 Ibid.
26 Niels K. Dyrlund personal communication.

27 Tage Kaarsted, *Admiralen: Andreas de Richelieu: Forretningsmand og politiker i Siam og Danmark* (Copenhagen, Odense Universitetsforlag, 1990), quoted during personal communication with Niels K. Dyrlund.

28 Ole Lange interview.

29 Ramsay, "Development of a Bureaucratic Polity," 138.

30 In the mid-1700s, the first known Danish collection of Thai plant specimens was made by a Danish student of Linnaeus, the surgeon and naturalist Johann Gerhard Koenig. (This collection was lost at sea.) In the 1860s, the Danish Botanical Society funded an expedition that resulted in four hundrd plant specimens being delivered to the herbarium in Copenhagen. By the end of the nineteenth century, however, with East Asiatic's logging operations well underway, "it was natural that the company should fund intensive field work in Thailand." In 1899 East Asiatic and the Carlsberg Foundation financed a major biological expedition in and around Ko Chang that resulted in thirteen hundred species being brought back to Europe. Among Danish plant collectors in Thailand, two other East Asiatic employees collected flora: Erik Lindhard collected eighty-one specimens while supervising logging in northern Thailand between 1901 and 1904 and A. P. N. Vesterdal collected five hundred plants during the period 1936–39. See Royal Danish Ministry of Education, *Thai-Danish Relations: 30 Cycles of Friendship* (Copenhagen, n.d.), 189–91.

31 East Asiatic, *Two Kingdoms*, 44; interview with Henrik Keiding, December, 1999.

32 Danish Ministry, *Thai-Danish Relations*, 192.

33 Interview with Kirsten Olesen, Carl Syrach-Larsen's daughter, January 2001.

34 Carl Syrach-Larsen, *Genetics in Silviculture*, trans. Mark L. Anderson, (Edinburgh: Oliver and Boyd, 1956).

35 Carl Syrach-Larsen, "Genetics in Teak (*Tectona grandis L.*)," *Royal Veterinary and Agricultural College Yearbook 1966* (Copenhagen, 1966), 234.

36 Henrik Keiding and Erik Kjær, "Teak-en træagtig staude med et stort potentiale: Historien om Dansk-Thailandsk samarbeijde om forædling og bevaring af teaktræer gennem 40 år," *Dansk Skovbrugs Tidsskrift* (December 1998): 137.

37 East Asiatic, *Two Kingdoms*, 44.

38 Jens Granhof interview.

39 Keiding and Kjær, "Teak," 131.

40 As of January 2004, the Danida Forest Seed Centre was integrated into the new Danish Centre for Forest Landscape Planning under Niels Elers Koch.

41 *DFSC Programme 1995/1999* (Humlebæk: Danida Forest Seed Centre, 1995).

42 Niels Elers Koch interview.

43 Keiding and Kjær, "Teak," 137.

44 Interview with Erik Kjær of the Danida Forest Seed Centre, February 2000.

45 Ibid.

46 Pinkaew Laungaramsri, "Danced's [Danish Cooperation for Environment and Development] Outdated Approach to Thai Environmental Problems," *Development Today* 21 (1999): 11.

47 Syrach-Larsen, "Genetics in Teak," 243.

48 Personal communication with Verapong Suangtho of the Royal Forestry Department, March 2000; Apichart Kaosa-ard interview; Jorni Odechao interview.

49 Long, "Displacement of Taungya Plantations," 281–82.

50 P. Chr. Nielsen, Kirsten Olesen, and Harry Petersen, eds., *Carl Syrach-Larsen, 6 Juli 1898–20 Januar 1979* (Copenhagen: Mindehefte undgivet af Skovhistorisk Selskab 1982), 11.

51 *The Empire Forestry Review* 1 (1948): 44, quoted in Ibid. This is my English translation from the Danish.

52 Nielsen, *Carl Syrach-Larsen*, 7.

53 Syrach-Larsen, *Genetics in Silviculture*, x, 2–3.

54 Ibid. (emphasis in the original)

55 Kirsten Olesen interview.

56 Jens Granhof interview.

57 Syrach-Larsen, "Genetics in Teak," 237. (Emphasis in the original.)

58 Keiding and Kjær, "Teak," 127.

59 Apichart Kaosa-ard interview.

60 Apichart Kaosa-ard, *Experience from Tree Improvement of Teak in Thailand: Technical Note 50* (Humlebæk: Danida Forest Seed Centre, 1998), 26.

61 Apichart Kaosa-ard interview.

62 The Forestry Industry Organization (FIO) manages about 800 sq km (500,000 rai) of teak plantations that were planted during the period from the early 1960s up until the logging ban. Interview with Manoonsak Tantiwat, managing director of FIO, January 2008.

63 Apichart Kaosa-ard, et al., "Teak," 41.

64 Ibid.

65 Apichart Kaosa-ard interview.

66 Apichart Kaosa-ard, *Experience from Tree Improvement*, 25.

67 H. Wellendork and Apichart Kaosa-ard, "Teak Improvement Strategy in Thailand," *Tree Improvement* 21 (1988): 1–43, quoted in Ibid., 25.

68 Jens Granhof interview.

69 Syrach-Larsen, *Genetics in Silviculture*, x.

70 Another planned water diversion project, the Kok-Ing-Nan, would inundate the Mae Yom forest by diverting water from the Mekong River through the Kaeng Suea Ten dam.
71 Niels Elers Koch interview.
72 Ibid.
73 Interviews with Karen villagers in Ban Wat Chan, February 1992.
74 According to M.R. Pattarachai Rachani of the Royal Forestry Department, natural pine was considered by foresters to be "value-less" wood until the early 1980s. Yet the prospect of setting up a sawmill in the forest sparked interest among many Japanese and Westerners looking to buy the wood.
75 The Logging Division, precursor of the Forestry Industry Organization (FIO), was established within the Royal Forestry Department in 1912 and given logging concessions in Mae Chaem. See Chamaichome Sunthornsawasdi, *Historical Study of Forestry*, 40.
76 Ibid., 49.
77 Follow-up interviews with Karen villagers in Ban Wat Chan, March 1992.
78 Dusit, "Forest Policy," 59.
79 Ibid., 60.
80 Jens Granhof interview.
81 Danish foresters working at the Pine Improvement Center during its first decade included Jens Granhof, who had come to Thailand as an FAO consultant and now oversees Forgenmap; and Palle Havmøller, whose father worked for the Danish East Asiatic Company. Havmøller subsequently became chief technical advisor for the Danish-funded Western Forest Complex project.
82 Bhargava Consulting and Design Engineers (P) Ltd., "Potential for Pulp and Paper Development—Thailand," in *FAO Pulp and Paper Industries Development Programme (Phase II)*. Rome: Food and Agricultural Organization, 1976. The study also identified bamboo, bagasse, and rice straw as suitable raw material for pulp but, interestingly, not eucalyptus.
83 Lert Chuntanaparp, et al., "Early Results of a Provenance Trial of *P. kesiya* in Thailand," in Sanga Sabhasri, *The Thai-Danish Pine Project 1969–1974* (Bangkok, n.d., ca. 1974), 23.
84 Ibid., 29.
85 Jens Granhof interview.
86 Ibid.
87 Interview with Reungchai Pao-sujja, head of Royal Forestry Department's Private Reforestation Promotion Office, 1991.
88 Reungchai Pao-sujja interview, March 1992.
89 Jens Granhof interview.

90 Interview with Suntisuk Prasitsak, project officer for Ban Wat Chan Project, Forestry Industry Organization (FIO), March 1992.

91 This section is based on interviews with Ban Wat Chan villagers conducted in March 1992. Fearing for their safety, they spoke on condition of anonymity.

92 Ban Wat Chan interviews, March 1992.

93 Interview with M.R. Adulyadej Chakraphand, director of FIO, March 1992.

94 Jaakko Pöyry Oy, "Ban Wat Chan Integrated Rural Development Project: Main Report (Nordic Investment Bank)," Helsinki, December 16, 1991.

95 Ibid.

96 Suntisuk interview.

97 The late former managing director of the Forestry Industry Organization (FIO), Amnuai Corvanich, quoted in Banasopit Mekvichai, "The Teak Industry," 243.

98 *Progress Report of Forest Administration in British Burma*, 1863–64, quoted in Raymond Bryant, "Shifting the Cultivator: The Politics of Teak Regeneration in Colonial Burma," *Modern Asian Studies* 28, no. 2 (1994): 234.

99 Interview with M.R. Adulyadej Chakraphand, October 1999.

100 Murray Bail, *Eucalyptus: A Novel* (Toronto: Vintage Canada, 1999), 15.

101 P. O. Olesen, "Notes from the 12th Inter-Nordic Course on Forests and Forestry in Developing Countries," Helsinki, September, 17–28 1984, 93.

102 J. W. Turnbull and J. C. Doran, "Role of the CSIRO Tree Seed Centre in Collection, Distribution and Improved Use of Genetic Resources of Australian Trees [International Union of Forest Research Organizations, Project Group: Paper No. 46]" (paper presented at the 12th Inter-Nordic Course on Forests and Forestry in Developing Countries, Helsinki, September 17–28, 1984), 2.04.00.

103 *Camaldulensis* is the second most common species of eucalyptus used in plantations worldwide after *Eucalyptus grandis*. See Evans, *Plantation Forestry*, 36.

104 Boonwong Thaiutsa, "Mai eucalyptus" (unpublished paper, 1990), 5. The only soil type that seems to inhibit the growth of *E. camaldulensis* in Thailand is alkaline.

105 Evans, *Plantation Forestry*, 339. *Camaldulensis* and other eucalypt species have been used to dry out marshy ground in countries as far-flung as Uganda, Italy, and Israel.

106 Pongsak Sahunalu, Chakraphol Chakrapholwareut, Pitaya Petmak, and Preecha Dhammanon, "Phon khong khwam na naen khong kan pluk pa to phon phalit khong suan pa eucalyptus camaldulensis thi pluk phuea kan prayuk rabap wanakaset," *Thai Journal of Forestry* 6 (1987): 213–38. In Thai.

107 Ibid.

108 Boonwong Thaiutsa, "Adit, patchuban lae anakhot khong mai eucalyptus nai prathet thai," *Sak Thong* 12, no. 4 (1986): 6–13. In Thai.

109 Boonwong Thaiutsa, "Mai eucalyptus."

110 Reaung Chai Pousugg, "Early Results of Species Trial of Eucalyptus in Thailand," in Sanga, *The Thai-Danish Pine Project*.
111 Danida Forest Seed Centre. *Pre-Appraisal Report: Forest Seed Supply, Tree Improvement and Gene Conservation in Thailand 1994–1997*. Annex 1 [Historical Background] (Humlebæk, July 1994).
112 Jens Granhof interview.
113 Ricardo Carrere and Larry Lohmann, *Pulping the South: Industrial Tree Plantations and the World Economy* (London: Zed Books, 1996), 233.
114 Boonwong Thaiutsa, "Mai eucalyptus," 2.
115 Jens Granhof interview.
116 Suda Kanjanawanana, "Eucalyptus protests going straight to the top," *The Nation*, March 30, 1989.
117 Sanitsuda Ekachai, "A life broken by plantations," *Bangkok Post*, February 27, 1991.
118 Boonwong Thaiutsa, "Mai eucalyptus," 3.
119 Ann Danaiya Usher, "After the forest . . ." *The Nation*, June 1990.
120 Boonwong Thaiutsa, "Mai eucalyptus," 3.
121 The seminar took place on June 19–21, 1987.
122 Boonwong Thaiutsa, "Mai eucalyptus," 4.
123 Ibid.
124 Interview with Boonwong Thaiutsa, Thai forestry professor, June 1990.
125 Duncan Poore and C. Fries, "The Ecological Effects of Eucalyptus." *Forestry Paper Series, No. 59* (Rome: Food and Agriculture Organization, 1985); see also the summary of this report, *The Eucalypt Dilemma* (Rome: Food and Agriculture Organization, 1988).
126 Interview with forester Pitaya Petmak, January 1990, in Ann Danaiya Usher, "The most (mis)quoted forester," *The Nation*, February 1990.
127 See Pitaya Petmak, Bopit Kietvuttinon, and Boonchoob Boontawee, *Some Ecological Impacts of Planting Eucalyptus* (Bangkok: Silvicultural Research Sub-division, Royal Forestry Department, 1986).
128 W. Jirasuktaveekul and P. Witthawatchutikul, *Soil Moisture of Eucalyptus Plantation and Abandoned Area at Ban Huay Ma Fuang, Rayong* (Bangkok: Research Section, Watershed Management Division, Royal Forestry Department, 1988).
129 I. Craig, S. Wasunan, and M. Saenlao, "Effects of Paddy-Bund-Planted Eucalyptus Trees on the Performance of Field Crops" (paper presented at the Fifth Annual Farming Systems Conference, Kamphaengsaen, 1988).
130 C. Homchand, S. Mongkholsawad, and T. Tulaphitak, *Impact of Eucalyptus Plantation on Soil Properties and Subsequent Cropping* (Bangkok: United States Agency for International Development, 1989).

131 Ron Aurell and Jaakko Pöyry, "Pulp and Paper: Worldwide Trends in Production, Consumption and Manufacturing," in *Global Issues and Outlook in Pulp and Paper*, ed. Gerard F. Schreuder (Seattle: University of Washington Press, 1987), 10.

132 Banasopit Mekvichai, "Teak Industry," 77.

133 Ann Danaiya Usher, "The tree farm that never was," *The Nation*, June 1990.

134 Martin Bayliss, "Economic Boom Boosts Paper Demand," *Pulp and Paper International* (April 1990): 38.

135 Patricia Marchak, "For Whom the Tree Falls: Restructuring of the Global Forest Industry" (paper presented to the joint meetings of the Canadian Political Science/Canadian Sociology and Anthropology Associations, Victoria, Canada, July 1990).

136 Aurell and Pöyry, "Pulp and Paper," 10.

137 Ibid.

138 Jaakko Pöyry, "Asia's Widening Fibre Deficit," *Pulp and Paper International* 31, no. 9 (1989): 5.

139 That is, until the year 2103. See, e.g., Arbhabhirama, et al., *Thailand Natural Resources Profile*.

140 10 baht per rai per year.

141 Thailand Development Research Institute, "Economic Forests: Myth or Reality?," *TDRI Quarterly Review* (Bangkok, 1989).

142 Ann Danaiya Usher, "The dilemma of the 1990s: eucalyptus vs. land rights," *The Nation*, May 5, 1989.

143 Ibid.

144 Boonwong Thaiutsa, "Mai eucalyptus," 3.

145 Reungchai Pao-sujja interview, March 1992.

146 Interview with Manote Winaipanich of the plantation promotion division of the Thai Pulp and Paper Association, October 1999.

147 Keith Barney, "At the Supply Edge: Thailand's Forest Policies, Plantation Sector and Commodity Export Links with China," in *China and Forest Trade in the Asia-Pacific Region: Implications for Forests and Livelihoods* (Washington, DC: Forest Trends, 2005), 13; interview with Permsak Mukarabhirom, January 15, 2008.

148 Tongroj Onchan claims to be responsible for quoting the estimate of 10 million people living illegally in forest reserve land, which was widely used in the forestry policy debate. In fact, there is no exact figure.

149 Article 16, 1964 National Forest Reserve Act.

150 Ann Danaiya Usher, "Eucalyptus widening the gap," *The Nation*, June 1990.

151 Thailand Development Research Institute, "Eucalyptus: For Whom and For What?," *TDRI Quarterly Review* (Bangkok, June 1990).

152 Ibid.

153 S. Saowakontha, et al., "Roles of Food Gathered from the Forest in Self-Reliance and Nutritional Status of Villagers in Northeast Thailand: The Case of Ban Non Khong," in *Why Natural Forests are Linked with Nutrition, Health and Self-Reliance of Villagers in Northeast Thailand*, ed. W. Brinkman (Khon Kaen: Khon Kaen University, 1989).

154 Suda Kanjanawanawan, "A chronology of protests," in "Eucalyptus protests going straight to the top," *The Nation*, March 30, 1989.

155 Interview with forestry consultant, February 1990. When the eucalyptus farms of Aracruz were first established in the 1970s, the Brazilian state of Espirito Santo was under military rule. Interview with Jose Augusto Padua of Greenpeace Brazil, February 1990.

156 Interview with Kamnan Viset Sinet, village leader in Yasothon Province, August 1992.

157 Krishna Brikshavana interview.

158 Thai NGO Committee, "Programme for Agricultural Land Distribution."

159 Plodprasob Suraswadi interview.

160 Upai Wayupat interview.

161 Jens Granhof interview.

162 Ann Danaiya Usher, "The tree farm that never was," *The Nation*, June 1990.

163 John Pearson, "Shell Keeps Going Well Into Forestry," *Pulp and Paper International* 30, no. 11 (November 1988), 35.

164 Ann Danaiya Usher, "Shell caught in eucalyptus debate: wonder tree or ecological menace?," *The Nation*, February 8, 1989.

165 Suan Kitti also quoted Phitaya Petmak's research as proof of the ecological benefits of large-scale eucalyptus plantations. In fact, his studies concluded nothing of the sort.

166 Ann Danaiya Usher, "Shell caught in eucalyptus debate: wonder tree or ecological menace?," *The Nation*, February 8, 1989.

167 It was this 2,000 rai limit that was reduced to 50 rai after the Suan Kitti scandal. It was subsequently doubled, but remains an important obstacle to the expansion of industrial tree farms in Thailand.

168 "Shell's multi-billion baht reforestation concession," *Appropriate Technology Association Magazine* 7, no. 2 (1988): 23.

169 "Eucalyptus project faces stiff opposition: Shell Group vows to carry out plan," *Bangkok Post*, December 28, 1987.

170 Tunya Sukpanich, "Villagers tell of Forestry's 'scare tactics,'" *Bangkok Post*, May 1988.

171 Interview with Police Lieutenant Prathin Santiprapob, February 1990.

172 Caroline Sargent, *The Khun Song Plantation Project: A Socio-Economic and Environmental Analysis Towards the Establishment and Management of Plantations in Chanthaburi Province by Shell Companies of Thailand* (London: Forestry and Land Use Programme, International Institute for Environment and Development, 1990).

173 Ann Danaiya Usher, "The tree farm that never was," *The Nation*, June 1990.

174 Jens Granhof interview.

175 Interview with Reungchai Pao-sujja, head of Royal Forestry Department's Private Reforestation Promotion Office, June 1990.

176 Interview with Narong Supsuwan, managing director, Thai-Japan Reforestation and Wood Industry Co. Ltd. (TJR), June 1990.

177 "Eucalyptus Forestation in Thailand," *Tradescope* 8, no. 9, (September 1988).

178 Ann Danaiya Usher, "Pulp links, from Oz to Siam," *The Nation*, June 1990.

179 The "forest cooperative" idea is almost certainly inspired by the Scandinavian forest co-ops. These are well-known to the Danish consultants involved in the tree seed improvement program that Reungchai had worked with for so many years.

180 Tunya Sukpanich, "The eucalyptus question," *Bangkok Post*, February 16, 1990.

181 "Finnish firm studies Thai paper industry: suggests ways to cope with future demand," *The Nation*, May 16, 1986.

182 Ann Danaiya Usher, "The shaping of a master plan," *The Nation*, June 1990; "A Finn-ancial harvest," *The Nation*, February 10, 1991.

183 Interview with Nat Inthakan, head of Jaakko Pöyry Thailand Co. Ltd., 1990.

184 Chris Lang, *The Pulp Invasion: The International Pulp and Paper Industry in the Mekong Region* (Maldonado: World Rainforest Movement, 2002).

185 Plodprasob Suraswadi interview.

186 Noel Rajesh, "Thailand: Sino-Thai Eucalyptus Project Facing Opposition," *World Rainforest Movement Bulletin* 34 (May 2000).

187 *Bangkok Post*, "Venture to build B27b pulp mill," March 25, 1997; *The Nation*, March 25, 1997.

188 Plodprasob Suraswadi interview.

189 Upai Wayupat interview.

190 Interview with Olle Zackrisson, Swedish forest biologist, 1994.

191 H. H. Cramer, "On the Predisposition to Disorders of Middle European Forests," *Pflanzenschutz-Nachrichten* 37 (1984): 102, quoted in Chris Maser, *The Redesigned Forest* (San Pedro, CA: R. & E. Miles, 1988), 78.

192 R. Plochmann, "Forestry in the Federal Republic of Germany," *Hill Family Foundation Series* (Corvallis: School of Forestry, Oregon State University, 1968), 52, quoted in Maser, *Redesigned Forest*, 80.

193 Ibid., 71

194 Ibid., 69.

195 Hatzfeldt, "Forestry in Germany."

196 Siegfried Lewark, "The Radical Revision of the Forestry Curriculum at the Freiburg Forestry Faculty" (paper delivered at FAO Advisory Committee on Forestry Education (ACFE) Meeting in Chile, 1996.)

197 Siegfried Lewark interview. In Germany, traditions of ecological forestry did exist. *Dauerwald* (or permanent forestry) was especially active in the 1920s and 1930s. *Dauerwald* was based on the management of complex, multi-aged, and multi-species stands that were "shrewdly cultivated through recurrent selective harvesting and natural regeneration." But these ideas tended to be "ridiculed as unscientific and were sometimes even suppressed" by mainstream German forestry, which focused overwhelmingly on even-aged monocultures. Readers of German should see the Ph.D. thesis of Irene Seling, "Die Dauerwald bewegung in den Jahren zwischen 1880–1930—eine sozialhistorische Analyse" (Schriften aus dem Institut für Forstökonomie der Universitet Freiburg, Band 8, 1997).

198 Kennedy, et al., "Values, Beliefs," 22.

199 Olle Zackrisson interview.

PART 4. THE MAKING OF THAI WILDERNESS

1 Ramachandra Guha, "Two Phases of American Environmentalism: A Critical History," in *Decolonizing Knowledge*, ed. Frédérique Apffel–Marglin and Stephen Marglin (Clarendon Press, Oxford, 1996), 129.

2 Meridel LeSueur, "The Ancient People and the Newly Come," in *Ripening: Selected Work, 1927–1980* (Old Westbury, NY: The Feminist Press, 1982), 39.

3 Interview with Phairoj Suwannakorn, first director of the National Parks Division, June 1993.

4 Roderick Nash, *Wilderness and the American Mind*, 3rd ed. (New Haven: Yale University Press, 1982), 67.

5 Ibid., 74.

6 Two books that address this problem are Mark David Spence, *Dispossessing the Wilderness: Indian Removal and the Making of the National Parks* (New York, Oxford University Press, 1999); and Robert H. Keller and Michael F. Turek, *American Indians and National Parks* (Tucson: The University of Arizona, 1998). I rely heavily on these two studies for this section, and highly recommend both for anyone interested in understanding the political history of America conservation thinking.

7 George Catlin, *North American Indians: Being Letters and Notes on their Manners, Customs, and Conditions, Written During Eight Years' Travel Amongst the Wildest Tribes of Indians in North America*, 2 vols. (Philadelphia, 1913), vol. 1, 289, 292–3, quoted in Nash, *Wilderness and the American Mind*, 101. Nash notes that these

volumes were written in the 1830s, but were originally published in London in 1841.

8 Interview with Surachet Chettamart, Thai forestry professor, April 1993. Yosemite and Yellowstone were also models for the creation of national parks in countries of Africa. See Jonathan Adams and Thomas McShane, *The Myth of Wild Africa: Conservation Without Illusion* (Berkeley: University of California Press, Berkeley, 1996).

9 See Stephen Fox, *The American Conservation Movement: John Muir and His Legacy* (Madison: The University of Wisconsin Press, 1981).

10 Ibid., 6, in correspondence with Ralph Waldo Emerson.

11 Ibid., 7.

12 Ibid., 44.

13 John Muir, *The Mountains of California* (New York: Century, 1894), 93, quoted in Spence, *Dispossessing the Wilderness*, 109.

14 Susanna Hecht and Alexander Cockburn, *The Fate of the Forest: Developers, Destroyers and Defenders of the Amazon* (New York: Harper Perennial, 1990).

15 Keller and Turek, *American Indians and National Parks*, 20.

16 Fox, *American Conservation Movement*, 4.

17 Allin, *Politics of Wilderness Preservation*, 25.

18 Fox, *American Conservation Movement*, 4.

19 Spence, *Dispossessing the Wilderness*, 103.

20 Keller and Turek, *American Indians and National Parks*, 20–1.

21 Spence, *Dispossessing the Wilderness*, 104, 106.

22 Ibid., 124, 130.

23 Keller and Turek, *American Indians and National Parks*, 22, 25.

24 Stephen Pyne, *World Fire: The Culture of Fire on Earth* (New York: Holt, 1995).

25 In 1896 the US Supreme Court ruled in *Ward v. Race Horse* that the off-reservation rights guaranteed to Shoshone and Bannock in 1868 were no longer valid. As Mark Spence points out, this decision clarified the legal status for native use of Yellowstone, even though Indians continued to use the park surreptitiously for hunting, gathering, and spiritual purposes. See Spence, *Dispossessing the Wilderness*, 64.

26 Ibid., 61.

27 Allin, *Politics of Wilderness Preservation*, 27.

28 Gustavus C. Doane, "The Report of Gustavus C. Doane Upon the So-Called Yellowstone Expedition of 1870 to the Secretary of War," *Early History of Yellowstone National Park and Its Relation to National Park Policies* (Washington, DC: Government Printing Office, 1932), quoted in Spence, *Dispossessing the Wilderness*, 43.

29 Spence, *Dispossessing the Wilderness*, 56.
30 Susanna Hecht and Alexander Cockburn, *The Fate of the Forest: Developers, Destroyers and Defenders of the Amazon* (New York: Harper Perennial, 1990).
31 Philetus Norris, *Report Upon the Yellowstone National Park to the Secretary of the Interior* (Washington, DC: Government Printing Office, 1877), 10, quoted in Spence, *Dispossessing the Wilderness*, 60.
32 Spence, *Dispossessing the Wilderness*, 23.
33 Ibid., 64.
34 Hecht and Cockburn, *The Fate of the Forest*.
35 Fox, *American Conservation Movement*, 346–49.
36 Keller and Turek, *American Indians and National Parks*, 91.
37 Ibid.
38 Chief White Calf, US Senate Document No. 118, 35, quoted in Keller and Turek, *American Indians and National Parks*, 49.
39 "Requiescant in Pace," *Glacier National Park Newsletter*, Glacier National Park (n.d.)
40 Spence, *Dispossessing the Wilderness*, 94.
41 Act of June 10, 1896, 29, quoted in Keller and Turek, *American Indians and National Parks*, 50.
42 Spence, *Dispossessing the Wilderness*, 78.
43 "Interview with Dr. George C. Ruhle," Glacier National Park Archives, file 7/14/83.
44 Keller, and Turek, *American Indians and National Parks*, 51.
45 George Ruhle, "Dr. George C. Ruhle's Talk, Conference Training Hall, Glacier National Park, August 3, 1975," Glacier National Park Museum and Archives, 2.
46 Keller and Turek, *American Indians and National Parks*, 59.
47 Ibid., 60
48 Ruhle, "Dr. George C. Ruhle's Talk," 5.
49 Spence, *Dispossessing the Wilderness*, 94–5.
50 Allin, *Politics of Wilderness Preservation*, 82.
51 Spence, *Dispossessing the Wilderness*, 87.
52 Interview with Vicky Santana, 1994, quoted in Keller and Turek, *American Indians and National Parks*, 63.
53 Boonsong Lekagul, *Hak pa mai*.
54 Ploenpote Atthakor, *Bangkok Post*, December 15, 2000.
55 Personal communication with Nantiya Tangwisuttiji, reporter for *The Nation*.

56 While American economic and military assistance to Thailand amounted to over US$1 billion between 1951 and 1967, aid from the World Bank through 1967 had a cumulative total of US$269 million, and aid from all other countries reached only US$10 million in 1964. See David A. Wilson, *The United States and the Future of Thailand* (New York: Praeger Publishers, 1970), 144–45.

57 Chris Dixon, *The Thai Economy: Uneven Development and Internationalisation*, (London: Routledge, 1999), 71, 176.

58 Pasuk Phongpaichit and Chris Baker, *Thailand: Economy and Politics* (Oxford: Oxford University Press, 1997), 60–64.

59 See Banasopit Mekvichai, "The Teak Industry," 182, 186, 191.

60 Tongroj Onchan, et al., *A Land Policy Study* (Bangkok: Thailand Development Research Institute, 1990), 48.

61 See, e.g., Prayong Nettayarak, "Wiwatthanakan khong kanbukboek thidin thamkin nai khetpa phak thawan ok chiang nuea," in Jermak Pinthong, *Wiwatthanakan khong kanbukboek thidin thamkin nai khetpa* (Bangkok: Local Development Institute, 1991), 201.

62 Adams and McShane, *The Myth of Wild Africa*, 27.

63 Interview with Nadda Sriyabhaya, friend and protégé of Boonsong Lekagul, October 20, 1999.

64 Ibid.

65 Nadda Sriyabhaya, "Dr. Boonsong Lekagul, Who Lit the Candle/Light/Torch of Conservation in Thailand" (paper prepared for the funerary book of Boonsong Lekagul at the request of his family, February 1992), 8.

66 Boonsong Lekagul, *Hak pa mai*, 197–8.

67 Ibid.

68 H. Huth, "Yosemite: The Story of an Idea", *Sierra Club Bulletin* 33, no.3 (1947), quoted in Guha, "Two Phases of American Environmentalism," 124.

69 George C. Ruhle, "Advisory Report on a National Park System for Thailand 1959–60: A report prepared for the International Union for Conservation of Nature and Natural Resources and the American Committee for International Wild Life Protection," *Special Publication No. 17* (New York: American Committee for International Wild Life Protection, 1964).

70 Marcus Colchester, "Salvaging Nature: Indigenous Peoples and Protected Areas" (discussion document of the United Nations Research Institute for Social Development with the World Rainforest Movement and Worldwide Fund for Nature, 1994), quoted in K. B. Ghimire and M. P. Pimbert, *Social Change and Conservation* (London: Earthscan, 1997), 99.

71 L. M. Talbot, "Wilderness Overseas," *Sierra Club Bulletin* 42, no. 6 (1957), quoted in Guha, "Two Phases of American Environmentalism," 124.

72 Nash, *Wilderness and the American Mind*, 343.

73 John Seed speaking at Stein Valley Festival, British Columbia, October 1988.
74 Personal communication with James Charleton, US National Parks Service officer, February 2, 1999.
75 "Glacier National Park's First Chief Naturalist Dies," *Hungry Horse News* (December 15, 1994), 5.
76 "Requiescant in Pace."
77 Personal communication with Robert Milne, former chief of International Parks Affairs Division, US National Parks Service, August 3, 2000; "Glacier National Park's First Chief Naturalist Dies," 5.
78 Ruhle, "Advisory Report."
79 Ibid.
80 Nadda Sriyabhaya, "Dr. Boonsong Lekaguld;" Nadda Sriyabhaya interview.
81 Boonsong Lekagul, *Hak pa mai*.
82 Elinor Ostrom, *Governing the Commons: The Evolution of Institutions for Collective Action* (Cambridge: Cambridge University Press, 1990), 23.
83 Terrestrial national parks covered 35,000 sq km, and wildlife sanctuaries made up 29,000 sq km. The plan at the time was to add 42 parks with an additional area of 24,500 sq km and 22 new sanctuaries in 7,600 sq km of forest. See Peter Vandergeest, "Property Rights in Protected Areas: Obstacles to Community Involvement as a Solution in Thailand," *Environmental Conservation* 23, no. 3 (1996): 259–68.
84 According to statistics obtained from the National Parks Department dated January 3, 2008, there are 108 national parks extending over 58,000 sq km and 52 wildlife sanctuaries covering 36,000 sq km.
85 The department aims to gazette a total of 116,800 sq km (over 73 million rai) as national parks and wildlife sanctuaries, with 58,200 sq km (32 million rai) of forest classified as Class 1A and Class 1B as well as two watersheds. See Ibid.
86 Out of 233 countries listed, Thailand is one of seventeen developing countries (not including island and city states) with more than 20 percent of their territory gazetted as protected areas. In Asia, only Bhutan (26%), Brunei (38%), and Cambodia (22%) have set aside more than 20 percent for conservation purposes. See *World Database on Protected Areas*, UNEP World Conservation Monitoring Centre, January 2007 (http://www.UNEP-wcmc.org/wdpa/index.htm? and http://www.UNEP-wcmc.org/wdpa/stats.cfm~summary_tab).
87 Oliver Pye, "Change and Continuity in the Royal Forestry Department," in *Khor Jor Kor: Forest Politics in Thailand* (Bangkok: White Lotus, 2005).
88 Veerawat Dheeraprasart, "Until No Trees Remain: Illegal Logging in the Salween Forest," in *After the Logging Ban: Politics of Forest Management in Thailand*, ed. Noel Rajesh (Bangkok: Foundation for Ecological Recovery, 2005); see also *The Nation*, January 17, 1996; *Bangkok Post*, January 24, 2006; "Governor, 3 district

chiefs face disciplinary inquiry," *Bangkok Post*, March 26, 1998; and "Governor says he will not resign," *Bangkok Post*, March 31, 1998.

89 Pinkaew Laungaramsri, "The Politics of Nature Conservation in Thailand," in *After the Logging Ban*.

90 Dong Phayayen–Khao Yai Forest Complex, UNESCO, July 2005 (http://whc.unesco.org/en/list/590).

91 "Khao Yai road works 'destroying' the park," *Bangkok Post*, May 3, 2006.

92 Chalermsak Vanichsombat interview, January 2008.

93 *Wanasarn* 58, no. 1 (2000).

94 See, e.g., Ploenpote Atthakor, "Declaration of park destroys livelihood," *Bangkok Post*, August 20, 2000; Sanitsuda Ekachai, "Why the racial violence now?," *Bangkok Post*, August 24, 2000; and "Prejudice dictates our every action," *Bangkok Post*, September 21, 2000.

95 Larry Lohmann, "Forest Cleansing: Racial Oppression in Scientific Nature Conservation," *Corner House Briefing* 13 (1999).

REINVENTING THAI FORESTRY

1 Interview with Anan Kanjanapan, Thai anthropologist, January 2008.

2 Interview with Somsak Sukwong, founding director of the Regional Community Forest Training Center for Asia and the Pacific (RECOFTC), January 2008.

3 Kasetsart University Alumni Association, "Naeo thang kan phatthana kan borihan chatkan pa mai khong chat" (unpublished paper, July 2007). In Thai.

4 Sanitsuda Ekachai, "Forests clearly deserve better," *Bangkok Post,* May 28, 1997.

5 Sanitsuda Ekachai, "Making a career of raping forests," *Bangkok Post*, March 30, 2000.

6 Kennedy, et al., "Values, Beliefs," 15.

7 Interview with forester Komol Praekthong, January 2008.

BIBLIOGRAPHY

BOOKS, ARTICLES, AND PAPERS

Adams, Jonathan, and Thomas McShane. *The Myth of Wild Africa: Conservation Without Illusion*. Berkeley: University of California Press, 1996.

Allin, Craig W. *The Politics of Wilderness Preservation*. Westport, CT: Greenwood Press, 1982.

Aphorn. See Withaya Aphorn.

Aurell, Ron, and Jaakko Pöyry. "Pulp and Paper: Worldwide Trends in Production, Consumption and Manufacturing." In *Global Issues and Outlook in Pulp and Paper*, edited by Gerard F. Schreuder. Seattle: University of Washington Press, 1987.

Bail, Murray. *Eucalyptus: A Novel*. Toronto: Vintage Canada, 1999.

Banasopit Mekvichai. "The Teak Industry in North Thailand: The Role of a Natural Resource-Based Export Economy in Regional Development." Ph.D. diss., Cornell University, 1988.

Banijbatana, Dusit. "Forest Policy in Northern Thailand." In *Farmers in the Forest: Economic Development and Marginal Agriculture in Northern Thailand*, edited by Peter Kunstadter, E. C. Chapman, and Sanga Sabhasri. Honolulu: East-West Center, University Press of Hawai'i, 1978.

Bayliss, Martin. "Economic Boom Boosts Paper Demand." *Pulp and Paper International* 32, no. 3 (April 1990): 38–41.

Boonsong Lekagul. *Hak pa mai yang yu yang yuen yong*, 1959. In Thai.

Boonwong Thaiutsa. "Adit, patchuban lae anakhot khong mai eucalyptus nai prathet thai." *Sak Thong* 12, no. 4 (1986): 6–13. In Thai.

———. "Mai Eucalyptus" [The Eucalyptus Tree]. Unpublished paper, 1990. In Thai.

Borwornsak Uwanno, and Wayne D. Burns. "The Thai Constitution of 1997: Sources and Process." *University of British Columbia Law Review* 32, no. 2 (1998): 227–33.

Bourke-Borrowes, D. "The Teak Industry of Siam." In *Technical and Scientific Supplement to the Record*. Bangkok: Ministry of Commerce and Communications, 1927.

———. "General Thoughts and Observations on Forestry in Siam." *Indian Forester* 54, no. 3 (1928): 141–60.

Brandis, Dietrich. *Notes on Forest Management in Germany*. Calcutta: Office of the Superintendent of Government Printing, India, 1866. Office of India Records, 1866. V/27/560/136.

———. *Suggestions Regarding Forest Administration in British Burma, January 25, 1876.* Office of India Records, 1876. V/27/560/72.

———. "The Progress of Forestry in India." *Indian Forester* 10, no. 11 (1884): 508–10.

———. *Indian Forestry* (1897): 77. Quoted in K. Sivaramakrishnan, "The Politics of Fire and Forest Regeneration in Colonial Bengal." Special issue: South Asia, *Environment and History* 2, no. 2 (June 1996): 145–94.

Brenner, Verena, et al. "Thailand's Community Forest Bill: U-turn or Roundabout in Forest Policy?" SEFUT Working Paper No. 3, revised ed., University of Freiburg, January 1999.

Brown, Ian. *The Élite and the Economy in Siam, c. 1890–1920.* Singapore: Oxford University Press, 1988.

Bryant, F. Beadon. "Fire Conservancy in Burma." *Indian Forester* 33, no. 12 (1907): 537–49.

Bryant, Raymond. "From Laissez-faire to Scientific Forestry: Forest Management in Early Colonial Burma, 1826–1885." *Forest and Conservation History* 34, no. 4 (October 1994): 160–70.

———. "Shifting the Cultivator: The Politics of Teak Regeneration in Colonial Burma." *Modern Asian Studies* 28, no. 2 (1994): 225–50.

Carrere, Ricardo, and Larry Lohmann. *Pulping the South: Industrial Tree Plantations and the World Economy.* London: Zed Books, 1996.

Catlin, George. *North American Indians: Being Letters and Notes on their Manners, Customs, and Conditions, Written During Eight Years' Travel Amongst the Wildest Tribes of Indians in North America*, 2 vols., Philadelphia, 1913. Quoted in Roderick Nash, *Wilderness and the American Mind*, 3rd ed. New Haven: Yale University Press, 1982.

Chainarong Setthachuea, and Malee Traisawasdichai. "Evicted." *The Nation*, January 22, 1992.

Chaiyan Rajchagool. *The Rise and Fall of the Thai Absolute Monarchy: Foundations of the Modern Thai State from Feudalism to Peripheral Capitalism.* Bangkok: White Lotus, 1994.

Chamaichome Sunthornsawasdi. *An Historical Study of Forestry in Northern Thailand Between 1896 and 1932.* Bangkok, 1978.

Chapin, Mac. "A Challenge to Conservationists." *World Watch Magazine* 17, no. 6 (November–December 2004). http://www.worldwatch.org/node/565.

Clawson, Marion, and Roger Sedjo. "History of Sustained-Yield Concept and Its Application to Developing Countries." In *History of Sustained Yield Forestry: A Symposium*, edited by Harold K. Steen. Durham: Forest History Society, 1984.

Colchester, Marcus. "Salvaging Nature: Indigenous Peoples and Protected Areas." Discussion document of the United Nations Research Institute for Social Development with the World Rainforest Movement and Worldwide Fund for Nature, 1994. Quoted in K. B. Ghimire and M. P. Pimbert, *Social Change and Conservation*. London: Earthscan, 1997.

Cox, Belinda Stewart. "Thailand's Nam Choan Dam—A Disaster in the Making?" *The Ecologist* 17, no. 6 (1987): 212–19.

Craig, I., S. Wasunan, and M. Saenlao "Effects of Paddy-Bund-Planted Eucalyptus Trees on the Performance of Field Crops." Paper presented at the Fifth Annual Farming Systems Conference, Kamphaengsaen, 1988.

Cramer, H. H. "On the Predisposition to Disorders of Middle European Forests." *Pflanzenschutz-Nachrichten* 37 (1984). Quoted in Chris Maser, *The Redesigned Forest*. San Pedro, CA: R. & E. Miles, 1988.

Dawkins, H. Colyear, and Michael S. Philip. *Tropical Moist Forest Silviculture and Management: A History of Success and Failure*. Oxon: CAB International, Wallingford, 1998.

Dixon, Chris. *The Thai Economy: Uneven Development and Internationalisation*. London: Routledge, 1999.

East Asiatic (Thailand) Public Co. Ltd. *A Tale of Two Kingdoms*. Bangkok, 1995.

Evans, Julian. *Plantation Forestry in the Tropics: Tree Planting for Industrial, Social, Environmental and Agroforestry Purposes*. Oxford: Clarendon Press, 1992.

"Eucalyptus Forestation in Thailand." *Tradescope* 8, no. 9 (Sept. 1988), published by Japan External Trade Organization (JETRO).

Feeney, D. "Agricultural Expansion and Forest Depletion in Thailand, 1900–1975." In *World Deforestation in the Twentieth Century*, edited by J. Richards and R. Tucker. Durham and London: Duke University Press, 1988.

Fernandez, E. E. *Rough Draft of a Manual of Indian Silviculture*. Dehra Dun: Indian Forest Service, 1891.

Fernow, Bernard. *The Economics of Forestry*. New York: Thomas Y. Crowell, 1902.

Fisher, W. R. "Review on the New Edition of Volume IV of Dr. Schlich's *Manual of Forestry: Forest Protection*." *Indian Forester* 33, no. 8 (1907): 351–53.

———. *Schlich's Manual of Forestry*. Vol. 4, *Forest Protection*. Translated and adapted from "Der Forstschutz" by Dr. Richard Hess. London: Agnew & Co. Ltd., 1907.

Fox, Stephen. *The American Conservation Movement: John Muir and His Legacy*. Madison: The University of Wisconsin Press, 1981.

"Glacier National Park's First Chief Naturalist Dies." *Hungry Horse News* (December 15, 1994). Glacier National Parks Archives.

Guha, Ramachandra. *The Unquiet Woods: Ecological Change and Peasant Resistance in the Himalaya*. Delhi: Oxford University Press, 1989.

———. "An Early Environment Debate: The Making of the 1878 Indian Forest Act." *Indian Economic and Social History Review* 27, no. 1 (January–March 1990): 65–84.

———. "Early Environmentalists in India: Some Historical Precursors." Foundation Day Lecture, Society for Promotion of Wastelands Development, New Delhi, June 15, 1993.

———. "Two Phases of American Environmentalism: A Critical History." In *Decolonizing Knowledge: From Development to Dialogue*, edited by Frédérique Apffel-Marglin and Stephen Marglin. Oxford: Clarendon Press, 1996.

Hatzfeldt, Hermann Graf. "Forestry in Germany: Old and New." Unpublished paper, 1996.

Hecht, Susanna, and Alexander Cockburn. *The Fate of the Forest: Developers, Destroyers and Defenders of the Amazon*. New York: Harper Perennial, 1990.

Heske, Franz. *German Forestry*. New Haven: Yale University Press, 1938.

Holm, David F. "A History of the Teak Industry in Thailand." Unpublished paper, Yale University, May 1969.

Huth, H. "Yosemite: The Story of an Idea." *Sierra Club Bulletin* 33, no. 3 (1947). Quoted in Guha, "Two Phases of American Environmentalism."

"Interview with Dr. George C. Ruhle." Glacier National Park Archives. 7/14/83.

Kaarsted, Tage. *Admiralen: Andreas de Richelieu: Forretningsmand og politiker i Siam og Danmark*. Copenhagen: Odense Universitetsforlag, 1990.

Kamala Tiyavanich. *Forest Recollections. Wandering Monks in Twentieth-Century Thailand*. Chiang Mai: Silkworm Books, 1997.

Keiding, Henrik, and Erik Kjær. "Teak - en træagtig staude med et stort potentiale: Historien om Dansk-Thailandsk samarbeijde om forædling og bevaring af teaktræer gennem 40 år." *Dansk Skovbrugs Tidsskrift* (December 1998): 125–41. In Danish.

Keller, Robert H., and Michael F. Turek. *American Indians and National Parks*. Tucson: The University of Arizona Press, 1998.

Kennedy, James J., Michael P. Dombeck, and Niels E. Koch. "Values, Beliefs and Management of Public Forest in the Western World at the Close of the Twentieth Century." In *New Requirements for University Education in Forestry: Proceedings of a Workshop Held in Wageningen, The Netherlands, July 30–August 2, 1997*, edited by P. Schmidt, J. Huss, S. Lewark, D. Pettenella, and O. N. Sastamoinen. Korbeek-Dijle: Drukkerij De Weide, 1998.

King, Hamilton. "Teak Industry of Siam." *USCR* 66 (July 1901): 305–10.

Koch, Niels Elers, and James J. Kennedy. "Multiple-Use Forestry for Social Values." *Ambio* 20, no. 7 (1991): 330–33.

Lange, Ole. *Den hvide elefanten: H. N. Andersens eventyr og ØK 1852–1914*. Copenhagen: Gyldendal, 1986.

———. *Jorden er ikke større . . . H. N. Andersen, ØK og storpolitikken 1914–37*. Copenhagen: Gyldendal, 1988.

Langston, Nancy. *Forest Dreams, Forest Nightmares*. Seattle: University of Washington Press, 1995.

Laungaramsri. See Pinkaew Laungaramsri.

———. "The Politics of Nature Conservation in Thailand." In *After the Logging Ban: Politics of Forest Management in Thailand*, edited by Noel Rajesh. Bangkok: Foundation for Ecological Recovery, 2005.

Le May, Reginald, *An Asian Arcady: The Land and Peoples of Northern Siam*. Cambridge: W. Heffer & Sons Ltd., 1926.

Lekagul. See Boonsong Lekagul.

LeSueur, Meridel. *Ripening: Selected Work, 1927–1980*. Old Westbury, NY: The Feminist Press, 1982.

Lewark, Siegfried. "The Radical Revision of the Forestry Curriculum at the Freiburg Forestry Faculty." Unpublished paper delivered at FAO Advisory Committee on Forestry Education (ACFE) Meeting, Chile, 1996.

Lohmann, Larry. "Forest Cleansing: Racial Oppression in Scientific Nature Conservation." *Corner House Briefing* 13 (1999): 1–24.

Long, G. R. "Displacement of Taungya Plantations by Improvement Fellings." *Indian Forester* 33, no. 6 (1907): 281–82.

Lowood, Henry E. "The Calculating Forester: Quantification, Cameral Science, and the Emergence of Scientific Forestry Management in Germany." In *The Quantifying Spirit of the 18th Century*, edited by Tore Frängsmyr, J. L. Heilbron, and Robin E. Rider. Berkeley: University of California Press, 1990.

Mantel, Kurt. "History of the International Science of Forestry with Special Consideration of Central Europe: Literature, Training, and Research from the Earliest Beginnings to the Nineteenth Century." In *International Review of Forestry Research*, vol. 1, edited by John A. Romberger and Peitsa Mikola. New York: Academic Press, 1964.

Marchak, Patricia. "For Whom the Tree Falls: Restructuring of the Global Forest Industry." Paper presented to the joint meetings of the Canadian Political Science/Canadian Sociology and Anthropology Associations, Victoria, Canada, July 1990.

Mekvichai. See Banasopit Mekvichai.

Muir, John. *The Mountains of California*. New York: Century, 1894.

Nadda Sriyabhaya. "Dr. Boonsong Lekagul, Who Lit the Candle/Light/Torch of Conservation in Thailand." Paper prepared for the funerary book of Boonsong Lekagul at the request of his family, February 1992.

Nash, Roderick. *Wilderness and the American Mind*, 3rd ed. New Haven: Yale University Press, 1982.

Nielsen, P. Chr., Kirsten Olesen, and Harry Petersen, eds. *Carl Syrach-Larsen: 6 Juli 1898–20 Januar 1979*. Copenhagen: Mindehefte undgivet af Skovhistorisk Selskab, 1982.

Nutalai. See Prinya Nutalai.

Olesen, P. O. "Notes from the 12th Inter-Nordic Course on Forests and Forestry in Developing Countries." Helsinki, September 17–28, 1984.

Ostrom, Elinor. *Governing the Commons: The Evolution of Institutions for Collective Action*. Cambridge: Cambridge University Press, 1990.

Pasuk Phongpaichit, and Chris Baker. *Thailand: Economy and Politics*. Oxford: Oxford University Press, 1997.

Pearson, John. "Shell Keeps Going Well, Into Forestry." *Pulp and Paper International* 30, no. 11 (November 1988): 34–39.

Peluso, Nancy. *Rich Forests, Poor People: Resource Control and Resistance in Java*. Berkeley: University of California Press, 1994.

Peluso, Nancy Lee, and Peter Vandergeest. "Genealogies of Forest Law and Customary Rights in Indonesia, Malaysia and Thailand." Draft paper, 1999.

Permpongsacharoen. See Witoon Permpongsacharoen.

Phongpaichit. See Pasuk Phongpaichit.

Phumiphamorn. See Suree Phumiphamorn.

Pinchot, Gifford. *Breaking New Ground*. New York: Harcourt, Brace, & Co., 1974.

Pinkaew Laungaramsri . "DANCED's [Danish Cooperation for Environment and Development] Outdated Approach to Thai Environmental Problems." *Development Today* 21 (1999): 11.

Plochmann, R. "Forestry in the Federal Republic of Germany." *Hill Family Foundation Series* 6 (1968): 52.

Pointon, A. C. *Bombay-Burmah Trading Corporation Limited 1863–1963*. Southampton: The Millbrook Press Limited, 1964.

Pongsak Sahunalu, Chakraphol Chakrapholwareut, Pitaya Petmak, and Preecha Dhammanon. "Phon khong khwam na naen khong kan pluk pa to phon phalit khong suan pa eucalyptus camaldulensis thi pluk phuea kan prayuk rabap wanakaset." *Thai Journal of Forestry* 6 (1987): 213–38. In Thai.

Pöyry, Jaakko. "Asia's Widening Fibre Deficit." *Pulp and Paper International* 31, no. 9 (1989): 5.

Prinya Nutalai. "Earthquakes and the Nam Choan Dam." *The Ecologist* 17, no. 6 (1987): 223–25.

Pye, Oliver. *Khor Jor Kor: Forest Politics in Thailand*. Bangkok: White Lotus, 2005.

Pyne, Stephen. *World Fire: The Culture of Fire on Earth*. New York: Holt, 1995.

Rajchagool. See Chaiyan Rajchagool.

Rajesh, Noel. "Thailand: Sino-Thai Eucalyptus Project Facing Opposition." *World Rainforest Movement Bulletin* 34 (May 2000). http://www.wrm.org.uy/bulletin/34/Thailandia.htm.

———, ed. *After the Logging Ban: Politics of Forest Management in Thailand*. Bangkok: Foundation for Ecological Recovery, 2005.

Ramsay, James A. "The Development of a Bureaucratic Polity: The Case of Northern Thailand." Ph.D. diss., Cornell University, 1971.

Rao, Y. S. "Thailand's Teak Forests Today." *Asian Timber* (January–February 1982): 42–45.

"Requiescant in Pace." *Glacier National Park Newsletter*, n.d.

Ribbentrop, R. *Forests in British India*. Calcutta: Office of the Superintendent of Government Printing, India, 1900. Office of India Records, 1900. V/27/560/8.

Ruhle, George C. "Dr. George C. Ruhle's Talk." Conference Training Hall, Glacier National Park, August 3, 1975. Glacier National Park Museum and Archives.

Sahunalu. See Pongsak Sahunalu.

Saowakontha, S., et al. "Roles of Food Gathered from the Forest in Self-Reliance and Nutritional Status of Villagers in Northeast Thailand: The Case of Ban Non Khong." In *Why Natural Forests Are Linked with Nutrition, Health and Self-Reliance of Villagers in Northeast Thailand,* edited by W. Brinkman. Khon Kaen: Khon Kaen University, 1989.

Schlich, Wilhelm. *Schlich's Manual of Forestry*. Vol. 1, *Forest Policy in the British Empire*, 4th ed. (revised and expanded). London: Agnew & Co. Ld., 1922. Quoted in Office of India Records, 1922. V/27/561/2.

Scott, James C. *Seeing Like a State: How Certain Schemes to Improve the Human Condition Have Failed*. New Haven: Yale University Press, 1998.

Seling, Irene. "Die Dauerwald bewegung in den Jahren zwischen 1880–1930—eine sozialhistorische Analyse." Ph.D. diss, Schriften aus dem Institut für Forstökonomie der Universitet Freiburg, Band 8. Freiburg, 1997. In German.

Setthachuea. See Chainarong Setthachuea.

"Shell's Multi-Billion Baht Reforestation Concession," *Appropriate Technology Association Magazine* 7, no. 2 (1988): 23.

Shepherd, Gill, et al. "Rights, Tenure, Governance and a More Pro-poor Vision for Conservation: What Should We Be Aiming At?" Background paper for the conference *Towards a New Global Forest Agenda*, Stockholm, October 29, 2007.

Slade, Herbert. "Too Much Fire-Protection in Burma." *Indian Forester* 22, no. 5 (1896): 172–76.

———. "The Maihongson Forests in Siam." *Indian Forester* 27, no. 5 (1901): 476–84.

Smalley, William A. *Linguistic Diversity and National Unity: Language Ecology in Thailand*. Chicago: The University of Chicago Press, 1994.

Spence, Mark David. *Dispossessing the Wilderness: Indian Removal and the Making of the National Parks*. New York: Oxford University Press, 1999.

Sriyabhaya. See Nadda Sriyabhaya.

Stebbbing, E. P. *The Forests of India*. Vol. 1. London: John Lane The Bodley Head Ltd., 1921.

Sunthornsawasdi. See Chamaichome Sunthornsawasdi.

Suree Phumiphamorn. "Maisak mueang sayam: Chut kamnot krom pa mai." In *Nitiyasarn Silapawattanatham*. Mimeograph, n.d. In Thai.

Svalesen, Leif. *The Slave Ship Fredensborg*. Bloomington: Indiana University Press, 2000.

Syrach-Larsen, Carl. *Genetics in Silviculture*. Translated by Mark L. Anderson. Edinburgh: Oliver and Boyd, 1956.

———. "Genetics in Teak (*Tectona grandis L.*)." *Royal Veterinary and Agricultural College Yearbook 1966*. Copenhagen, 1966.

Talbot, L. M. "Wilderness Overseas." *Sierra Club Bulletin* 42, no. 6 (1957). Quoted in Guha, "Two Phases of American Environmentalism."

Thai NGO Committee on Forest and Land Problems in Northeast Thailand. "Programme for Agricultural Land Distribution for Poor People Living in Degraded Forest Reserves (*Khor Chor Kor*): Questions and Answers." Unpublished paper, 1992.

Thaiutsa. See Boonwong Thaiutsa.

Thompson, Virginia. *Thailand: The New Siam*. New York: The MacMillan Company, 1941.

Thongchai Winichakul. *Siam Mapped: A History of the Geo-body of a Nation*. Chiang Mai: Silkworm Books, 1994.

Tiyavanich. See Kamala Tiyavanich.

Tottenham, W. F. L. "The Formation of the Forest Department in Siam." *Indian Forester* 31, no. 8 (1905): 446–49.

———. "Forestry." In *The Kingdom of Siam 1904*, edited by A. Cecil Carter. Bangkok: The Siam Society, Bangkok, 1988.

Troup, R. S. *A Note on Some European Sylvicultural Systems, with Suggestions for Improvements in Indian Forest Management*. Calcutta: Office of the Superintendent Government Printing, India, 1916. Office of India Records, 1916. V/27/560/35.

———. *The Silviculture of Indian Trees*. Vol. 1. Oxford: Clarendon Press, 1921.

Turnbull, J. W., and J. C. Doran. "Role of the CSIRO Tree Seed Centre in Collection, Distribution and Improved Use of Genetic Resources of Australian Trees (International Union of Forest Research Organizations, Project Group Paper No. 46)." Presentation at the 12th Inter-Nordic Course on Forests and Forestry in Developing Countries, Helsinki, Finland, September 17–28, 1984.

Uwanno. See Borwornsak Uwanno.

Vandergeest, Peter. "Property Rights in Protected Areas: Obstacles to Community Involvement as a Solution in Thailand." *Environmental Conservation* 23, no. 3 (1996): 259–68.

———. *Development's Displacement: Ecologies, Economies, and Cultures at Risk*. Vancouver: University of British Columbia Press, 2007.

Walker, H. C. "Fire Protection in Burma." *Indian Forester* 34, no. 6 (1908): 339–49.

———. "Scientific Forestry." *Indian Forester* 33, no. 10 (1907): 452–56.

Wellendork, H., and Apichart Kaosa-ard. "Teak Improvement Strategy in Thailand." *Tree Improvement* 21 (1988): 1–43, quoted in Apichart Kaosa-ard, *Experience from Tree Improvement of Teak in Thailand: Technical Note 50*. Humlebæk: Danida Forest Seed Centre, 1998.

Wilson, Constance M. *Thailand: A Handbook of Historical Statistics*. Boston: G. K. Hall & Co., 1983.

Wilson, David A. *The United States and the Future of Thailand*. New York: Praeger Publishers, 1970.

Winichakul. See Thongchai Winichakul.

Winters, Robert. *The Forests and Man*. New York: Vantage Press, 1974.

Withaya Aphorn. "Kreuakhai khong paachumchon nai prathetthai pi 2004, Krongkarn Sitthichumchonseuksa." *Trang* (April 2004). In Thai.

Witoon Permpongsacharoen. "Tropical Forest Movements: Some Lessons from Thailand." In *Volume on Tropical Forests*, edited by Heffa Schücking. Sassenberg: Arbeit Gemeinschaft Regenwald 1991.

Witt, D. O. "The Use and Abuse of Forest Work in Siam." *Indian Forester* 30, no. 7 (1904): 299–303. Quoted in Pye, *Khor Jor Kor*.

REPORTS

Anat Arbhabhirama, et al. *Thailand Natural Resources Profile*. Bangkok: Thailand Development Research Institute, 1987.

Apichart Kaosa-ard. *Experience from Tree Improvement of Teak in Thailand: Technical Note 50*. Humlebæk: Danida Forest Seed Centre, 1998.

Apichart Kaosa-ard, Verapong Suangtho, and Erik Kjær. "Teak (*Tectona grandis Linn. F.*)" In *ASEAN: A Survey Report*. Bangkok: ASEAN/Canada Forest Seed Centre, October 1986.

Arbhabhirama. See Anat Arbhabhirama.

Barney, Keith. "At the Supply Edge: Thailand's Forest Policies, Plantation Sector and Commodity Export Links with China." In *China and Forest Trade in the Asia-Pacific Region: Implications for Forests and Livelihoods*. Washington, DC: Forest Trends, 2005.

Bhargava Consulting and Design Engineers (P) Ltd. "Potential for Pulp and Paper Development—Thailand." In FAO *Pulp and Paper Industries Development Programme (Phase II)*. Rome: Food and Agricultural Organization, 1976.

Boontawee, B. "Thailand Country Report on State of Forestry." Report presented at the 18th Session of the Asia-Pacific Forestry Commission, FAO. Noosaville, Queensland, Australia, May 15–19, 2000.

British Burmah Revenue and Agriculture Proceedings: Forests, January to June 1896. Part 11-B, file no. 2A7. January 1896.

Charoenniyophrai. See Udom Charoenniyophrai.

Chuntanaparp. See Lert Chuntanaparp.

Danida Forest Seed Centre. *Pre-Appraisal Report: Forest Seed Supply, Tree Improvement and Gene Conservation in Thailand 1994–1997.* Annex 1 (Historical Background). Humlebæk, July 1994.

———. *DFSC Programme 1995/1999.* Humlebæk, 1995.

Gustavus C. Doane, "The Report of Gustavus C. Doane Upon the So-Called Yellowstone Expedition of 1870 to the Secretary of War." In *Early History of Yellowstone National Park and Its Relation to National Park Policies.* Washington, DC: Government Printing Office, 1932. Quoted in Spence, *Dispossessing the Wilderness,.*

Homchand C., S. Mongkholsawad, and T. Tulaphitak. *Impact of Eucalyptus Plantation on Soil Properties and Subsequent Cropping.* Bangkok: United States Agency for International Development, 1989.

Jaakko Pöyry Oy. "Ban Wat Chan Integrated Rural Development Project. Main Report (Nordic Investment Bank)." Helsinki, December 16, 1991.

Janthawong. See Montree Janthawong.

Jirasuktaveekul, W., and P. Witthawatchutikul. *Soil Moisture Under Eucalyptus Plantation and Abandoned Area at Ban Huay Ma Fuang, Rayong.* Bangkok: Research Section, Watershed Management Division, Royal Forestry Department, 1988.

Kaosa-ard. See Apichart Kaosa-ard.

Kasetsart University Alumni Association. "Naeo thang kan phatthana kan borihan chatkan pa mai khong chat." Unpublished paper, Bangkok, July 2007. In Thai.

Lang, Chris. *The Pulp Invasion: The International Pulp and Paper Industry in the Mekong Region.* Maldonado: World Rainforest Movement, 2002.

Lert Chuntanaparp, et al. "Early Results of a Provenance Trial of *P. kesiya* in Thailand." In Sanga Sabhasri, *The Thai-Danish Pine Project 1969–1974.* Bangkok, n.d., ca. 1974.

Loetsch, F. *Report to the RTG on Forest Inventory of the Northern Teak Bearing Provinces.* Rome: Food and Agriculture Organization, 1958.

Ministry of Agriculture, Thailand. *Classification of Uses of Forest Land in National Forest Reserves.* Bangkok: National Forest Resource Management Unit of the National Forest Resource Lang Management Division, Royal Forestry Department, 1992.

Montree Janthawong. "Rai-ngan phon kan wichai rueang kwamlaklai thang chiwaphap lae rabop niwet nai khet pa chumchon phak nuea ton bon." Unpublished report, Chiang Mai, 1998. In Thai.

National Economic and Social Development Board and the United States Agency for International Development. *Safeguarding the Future: Restoration and Sustainable Development in the South of Thailand.* Bangkok, 1989.

Nelson, J. C. "Forest Settlement Report of the Garhwal District." Lucknow, 1916. Quoted in Guha, *The Unquiet Woods*.

Nettayarak. See Prayong Nettayarak.

Norris, Philetus. *Report Upon the Yellowstone National Park to the Secretary of the Interior.* Washington, DC: Government Printing Office, 1877. Quoted in Spence, *Dispossessing the Wilderness*.

Onchan. See Tongroj Onchan.

Pitaya Petmak, Bopit Kietvuttinon, and Boonchoob Boontawee. *Some Ecological Impacts of Planting Eucalyptus*. Bangkok: Silvicultural Research Sub-division, Royal Forestry Department, 1986.

Poore, Duncan, and C. Fries. "The Ecological Effects of Eucalyptus." *Forestry Paper Series, No. 59*. Rome: Food and Agriculture Organization, 1985.

Pousugg. See Reaung Chai Pousugg.

Prayong Nettayarak. "Wiwatthanakan khaung kanbukboek thidin thamkin nai khetpa phak thawanauk chiang neua." In *Wiwatthanakan khaung kanbukbeuk thidin thamkin nai khetpa*, edited by Jermak Pinthong. Bangkok: Local Development Institute, 1991. In Thai.

Progress Report of Forest Administration in British Burma, 1863–64. Quoted in Raymond Bryant, "Shifting the Cultivator: The Politics of Teak Regeneration in Colonial Burma," *Modern Asian Studies* 28, no. 2 (1994): 225–50.

Progress Report of Forest Administration in British Burma, 1865–66. Quoted in Raymond Bryant, "From Laissez-faire to Scientific Forestry: Forest Management in Early Colonial Burma, 1826–1885." *Forest and Conservation History* 38, no. 4 (October 1994): 160–70.

Reaung Chai Pousugg. "Early Results of Species Trial of Eucalyptus in Thailand." In Sanga Sabhasri, *The Thai-Danish Pine Project*.

Report on Forest Administration in Burma for 1922–23 (Glossary). Rangoon: Office of the Superintendent of Government Printing, Burma, 1923. Office of India Records, 1923. V/24/1397.

Royal Danish Ministry of Education. *Thai-Danish Relations: 30 Cycles of Friendship*. Copenhagen, n.d.

Royal Forestry Department. *Prawat krom pa mai 2439–2500* [The History of the Royal Forestry Department]. Bangkok, 1958. In Thai.

———. *Prawat krom pa mai, 2439–2514* [The History of the Royal Forestry Department]. Bangkok, 1971. In Thai.

———. *Forestry Statistics of Thailand 1985*. Bangkok: Planning Division, Royal Forestry Department, 1985.

———. *Forestry Statistics of Thailand 1998*. Bangkok: Data Center, Information Office, Royal Forestry Department, 1998.

———. *Forestry Statistics of Thailand 1999*. Bangkok: Royal Forestry Department, 1999.

Ruhle, George C. "Advisory Report on a National Park System for Thailand 1959–60: A Report Prepared for the International Union for Conservation of Nature and Natural Resources and the American Committee for International Wild Life Protection." *Special Publication, No. 17*. New York: American Committee for International Wild Life Protection, 1964.

Sabhasri. See Sanga Sabhasri.

Sanga Sabhasri. *The Thai-Danish Pine Project 1969–1974*. Bangkok, n.d., ca. 1974.

Sargent, Caroline. *The Khun Song Plantation Project: A Socio-Economic and Environmental Analysis Towards the Establishment and Management of Plantations in Chanthaburi Province by Shell Companies of Thailand*. London: Forestry and Land Use Programme, International Institute for Environment and Development, 1990.

Slade, Herbert. "Rai-ngan khong Mr. H. Slade rueang kan pa mai khong prathet sayam" [Report of Mr. H. Slade Concerning Forestry in Siam]. Reprint, Bangkok: Information Department of the Office of Forestry Secretariat, 1896. In Thai.

———. "Report of the Royal Forestry Department for 119 (1901)." National Archives of Thailand. File MR 5 M/41/1.

"Statement of the Case of the Estate of the Late Dr. Marion Alonzo Cheek Against the Siamese Government." National Archives Thailand. File MR 5 M/41/2.

Team Consulting Co. Ltd. *Environmental and Ecological Impacts*. Bangkok, 1980.

Thailand Development Research Institute. "Economic Forests: Myth or Reality?" *TDRI Quarterly Review*. Bangkok, 1989.

———. "Eucalyptus: For Whom and For What?" *TDRI Quarterly Review*. Bangkok, 1990.

Tongroj Onchan, et al. *A Land Policy Study*. Bangkok: Thailand Development Research Institute, 1990.

Udom Charoenniyophrai, ed. *Research Report on Indigenous Knowledge, Customary Use of Natural Resources and Sustainable Biodiversity Management: Case Study of Hmong and Karen Communities in Thailand*. Chiang Mai: Inter Mountain Peoples Education and Cultures in Thailand Association (IMPECT), June 2006.

INDEX

Adulyadet Chakraphand, 110
Agriculture Land Reform Office, 87
Ahka, 181
Amnuai Corvanich, 83
Anan Kanjanapan, 183, 184
Anand Panyarachun, 31, 131, 163
Andersen, Hans Nils, 64, 65, 92, 93–95
Anglo-Thai Company, 53, 65
Apichart Kaosa-ard, 78, 79, 97, 102, 103
Aracruz
   as model for eucalyptus plantations in Thailand, 26, 28
Australia
   British colonial forestry in, 38
   Burma's forest law use in, 56
   deep ecologist in, 169
   Jaakko Pöyry Oy plantations in, 137
   Japanese pulp production and imports from, 136
   opposition to industrial forestry in, 125, 136
   origins of eucalyptus, 113, 115

Ban Wat Chan, 105–12, 139
Banasopit Mekvichai, 71, 75, 77, 80
Bangkok Post, 19, 20
Bhichai Rattakul, 17, 21, 125
Bhumibol Hydroelectric Dam, 107, 171, 172
biological diversity, 12, 13, 14, 18, 187
   destruction of, 40
   German forests and, 38, 144
   reversal of Forestry Department mandate, 12, 73, 74, 183
Blackfeet Indians, 154, 157–60, 162–63
Bombay Burmah Trading Corporation, 53–54, 64–65

Slade's negotiations with, 59–60
   timber concessions canceled by Thai government, 95
Boonsong Lekagul, 61, 79, 163–73
Boonwong Thaiutsa, 115, 118
Borneo Company, 53, 54, 60, 63–65, 71
Borwornsak Uwanno, 4
Bourke-Borrowes, D., 75, 82
Bowring Treaty, 63
Brandis, Dietrich, 38–39, 43–44, 48–49
   on forest communities, 46–47, 69
   ideal forest and, 127
   Slade and 57, 85
Brandis system in Thailand, 78
   negative selection and, 79
Buddhism, 2

Castenskiold, Jørgen, 56, 63, 94–95
Catlin, George, 149, 150
Chaiyan Rajchagool, 67
Chalermsak Vanichsombat, 173, 179
Champion, H. G.
   on Syrach-Larsen, 100
Chatichai Choonhavan, 17, 20, 29, 72
Cheek, Marion Alonzo, 57, 63–64, 95
"Chipko" forests, 105
Choompol Ngamponsi, 13, 14, 16
Chulalongkorn, H.M. King, 37, 52, 53, 56, 57, 64, 92–95
colonial forestry, 39, 41, 58, 66–72, 100, 105
   antagonism with local people, 18
   awareness of human presence in Asian forests, 8
   Ban Wat Chan and, 109
   Brandis and, 43–44
   in Burma, 55

destruction of forest diversity and, 40, 74
fire suppression and, 49, 85
homogeneity and, 28
National Forest Reserves and, 61
resistance to, 47, 66–67, 70–72
Schlich's Manual of Forestry and, 45
taungya and, 83
colonial powers
British colonial ambitions in Siam, 51, 54
Danish commercial interests, 91
German colonial forestry, 41
importance of teak to, 37
Siam caught between British and French, 56
commercial tree farms. *See* plantation forestry
Community Forest Bill, 2, 3–5, 7, 25, 182, 184–86
constitutionality of, 7, 25, 186
conservation
advocates, 12, 17, 19, 32, 35, 61, 98–99, 172
approaches, 1, 3, 8, 187–88
American, 149, 150, 151, 152, 156–58, 163, 168–69
areas, 87, 125, 173, 174, 178, 179, 180, 181
attitude towards highlanders, 98, 151
laws, 10, 19, 151, 158, 170–71, 178, 185
mandate of Thai forestry administration, 6, 7, 73, 114, 147–48, 164, 168, 171, 180–81
versus logging and commercial forestry, 10, 33, 103, 173, 175, 183
"conservation imperialism," 147, 168
Constitution of the Kingdom of Thailand, 4–5, 24, 25, 164
of 1997 and 2007, 184
Constitutional Court, 5, 7, 25
corruption, 12, 33, 35, 74, 78–81, 151, 175

Damrong Pidej, 178–79
Damrong, Prince (Rajanupab), 51, 56, 57, 59–60, 62, 94
Danida Forest Seed Centre, 97, 98, 103
Danish East Asiatic Company, 53, 64–65, 91
botanical survey, 95, 96, 97, 100, 115
Castenskiold and, 95
connections with Danish elites, 94
founded by Hans Nils Andersen, 92–93
Danish influence on Thai forestry, 91–99
Decho Chayathap, 5

Electricity Generating Authority of Thailand (EGAT)
Nam Chon and, 13–16
on planned FAO pulp mill, 107
eucalyptus, 9, 12, 25–29, 43, 89, 107, 113–23, 143, 173
effect of diversity, 74, 90
promotion of, 91, 96, 124–28
protests against planting of, 129–31
pulp industrialists and, 132–42
threat to food security, 130
exotic tree species, 9, 96, 97, 127, 173, 182

FAO, 76, 97, 106–8, 115, 118–19
Fernow, Bernard, 41, 45
FIO. *See* Forestry Industry Organization
fire, use of, 2, 48, 105
analogy between eucalyptus and, 119
Brandis's thinking on, 47, 70
in colonial forestry, 89, 90
cover story headline of *Wanasan*, 181
forestry policy and, 126
in German forests, 143
pine and, 109, 110, 112
Shell and, 135
Slade's thinking on, 49–51, 85

teak and, 100, 103
  in Yellowstone, 154, 155
Food and Agriculture Organization.
  See FAO
Forest Genetic Conservation Management Program, 98
Forestry Industry Organization (FIO), 72, 73, 99, 103, 116, 165
  auction of confiscated logs, 80
  Ban Wat Chan and, 108–12
  forest village system, 84
  origins of, 62, 71
  planting of teak, 82
Forest Preservation Act, 67
Foundation for Ecological Recovery, 184
French East Asiatic Company, 53, 65, 105

German forestry, 8–9, 39–43
  "final demise" of, 143
  Freiburg University and, 41
  influence on Europe, Asia, and North America, 41
  purpose of, 42
  state control and, 38
  transfer to Asia, 43–44
Glacier National Park, 10, 157–63, 169, 180
Granhof, Jens, 98, 107, 108, 132
Great Northern Railway Company, 160
Guha, Ramachandra, 147

Hatzfeldt, Hermann, 143
Hmong, 181
homogeneity
  factory forests and, 90
  as ideal of forestry, 28
  US conservation movement and, 157
Huai Kaeo, 11, 22–25, 32
Huai Kha Khaeng Wildlife Sanctuary, 13, 18–20, 32–34
  UNESCO World Heritage status, 13, 19
Humphreys, Christmas, vi

illegal logging, 5, 35, 74, 79, 80, 84, 165, 170, 175
industrial plantations, 9, 123, 128, 187
  See also monoculture plantations; plantation forestry
International Commission on National Parks, 168
International Union for the Conservation of Nature (IUCN), 168, 169

Jaakko Pöyry Oy, 108, 126, 132
  Ban Wat Chan project, 108–12
  forestry master plan, 137–40
  See also Pöyry, Jaakko
Jorni Odechao, 68–69

Karen, 12, 13, 15, 68–89, 177, 178, 181
  of Ban Wat Chan, 105, 109–12
  folk song, 66
  Nam Chon and, 12, 16
  in Salween hill tracts, 49
Kasem Chunkao, 13, 16
Kasetsart Alumni Association, 184–85
Kasetsart University
  conservation debate held at, 19–20
  criticism of forestry training program, 167, 186
  faculty seminar, April 1990, 118
  survey of Nam Chon
    by university professors, 13
    assisted by university students, 15, 23
Keiding, Henrik, 98
Kennedy, Jim, 144, 187
Khao Yai National Park, 170–72
  dead baby elephants in, 176–77
  as World Heritage Site, 178–79
Khor Chor Kor, 12, 29–32, 131
  cancellation of, 31, 86, 131

ethnic minorities and, 181
resettlement of landless people by military force, 12, 30, 32, 86, 131, 181
RFD crisis due to, 173
Kitti Damnerncharnvanich, 26, 27, 28, 132, 137, 140–42
See also Suan Kitti Reforestation Company
Kjær, Erik, 98, 103
Koch, Niels Elers, 104, 145
Komol Praekthong, 186, 188
Kukrit Pramoj, 17, 33

Laitalainen, Rauno, 139
land tenure, 2, 9
1985 Forestry Policy, 128
Sino-Thai plantations project, 141
Le May, Reginald, 54, 58
Leonowens, Anna, 64
Leonowens, Louis, 64
LeSueur, Meridel, 148
Lewark, Siegfried, 43, 144, 187
Lincoln, Abraham, 150
logging ban, 17–22, 24, 62, 73, 77, 80, 86, 108, 110, 125, 127, 173, 175, 180
See also conservation: mandate of Thai forestry administration
logging concessions. *See* timber concessions
Long, G. R., 49
Louis Leonowens Company Ltd., 53, 64
Lua, 106

Mae Chan Ta, 15, 16
misanthropic forestry, 12, 35
monoculture plantations
bio-diversity depletion, 74
clone-based, 40
Danish shift away from, 97
degraded forest and, 128
popular resistance to, 3
risks of, 48
vulnerability, 90
See also industrial plantations; plantation forestry
Muir, John, 152, 168

Nam Chon dam, 11, 12–16, 17, 18, 33, 134, 172
Nash, Roderick, 149, 169
Nation, The (Bangkok), 11, 12, 27, 137
National Counter Corruption Commission, 23
National Forest Reserves, 4, 21, 24, 61, 86, 128, 136
degraded forest in, 126
National Forest Reserves Act, 61
national parks, 3, 7, 10, 149, 151, 163, 164, 168–72, 174, 176, 180
American, 10, 16, 147, 148, 150, 151, 156–58, 162, 168, 180
conflicts with people, 25, 151
See also Glacier; Yosemite; Yellowstone
National Parks, Wildlife and Plant Conservation Department, 6–7, 148, 173, 174, 178
Neah Teuh, 177–78
Nong Yai village, 30, 31
non-timber products, 25, 45, 72, 74, 98, 188
normal forests, 8, 42, 45, 73, 83, 89, 127
Norris, Philetus, 155–56
Northern Pacific Railway Company, 155
Nuan Lachai, 71–72

Oriental Hotel, 64, 91, 93
Ostrom, Elinor, 173

Phairoj Suwannakorn, 16, 23–24, 138
Pinchot, Gifford, 41
pine, 9, 90, 114
Ban Wat Chan and, 105–6, 109–12

biodiversity depletion and, 74
Central American pine, 115
compared with teak, 55
Danish role in identification of species, 96
efficiency compared with eucalyptus, 123
establishment of pine plantations 106–8
lodge-pole pine in Sweden, 145–46
Royal Dutch Shell Group and, 132
Scotch pine in Germany, 43
seed improvement, 91
Veerawat Dheeraprasart on, 89
Pine Improvement Center, 96–98, 106, 107, 108, 115
Pinkaew Laungaramsri, 99, 176
Pitaya Petmak, 115–17, 119–21, 123, 133
plantation forestry, 9, 10, 81–84, 91, 129
  1985 Forestry Policy and, 124–26, 127
  eucalyptus trials, 115
  incendiarism and, 70, 130
  green revolution forestry, 99
  Jaakko Pöyry Oy, 137–40
  large versus small scale, 129
  narrowing of species focus, 9
  risky ecological experiment, 143, 145
  Shell Thailand, 132–35
  state's failure with, 74, 80
  Swedish experience with, 145–46
  taungya, 69.
  See also Danish influence on Thai forestry; eucalyptus; industrial plantations; monoculture plantations
Plodprasob Suraswadi, 78, 141–42, 180–81
Pöyry, Jaakko, 123, 125
  See also Jaakko Pöyry Oy
Prem Tinsulanonda, 33, 139
Project for Ecological Recovery, 18

pulp industry
  eucalyptus and, 91, 113, 114, 115
  pine and, 106, 107
  shifting operations to warmer climates, 124
  strategies for acquiring land, 132–42
  Suan Kitti and, 27, 28, 140
    Aracruz, 26
  in United States, 125

racism, 88, 148, 157, 163, 181
Rapee Sakrit, 117
RECOFTC, 4, 86, 125, 183, 186
reforestation, 3, 12, 21, 27, 81, 127, 173
  1985 forestry policy, 125, 126, 182
  in Brazil, 130
  Huai Kaeo and, 22, 24
  in Java, 70
  Khor Chor Kor and, 29, 30, 32, 131
  monoculture stands, 127–28
  Suan Kitti and, 25, 27
Regional Community Forestry Training Center for Asia and the Pacific. See RECOFTC
Reungchai Pao-sujja, 97, 108, 115, 127, 136
Ribbentrop, R., 82
Richelieu, Andreas du Plessis de, 92, 94
Royal Forestry Department (RFD), 3, 5–7, 37, 39, 69, 173, 186
  arrest of director general, 175
  conservation forestry and, 147, 148, 164, 168, 170–71, 173, 174
  Danish financing of programs, 98
  establishment of, 56–59, 66
  eucalyptus and, 114, 115, 117, 119, 127
  Huai Kaeo and, 22, 24
  Kitti Damnerncharnvanich and, 26, 27, 141–42
  measures to deal with illegal logging, 80
  logging ban and, 17, 21

logging concessions and, 52, 55, 67
participation in *Khor Chor Kor*, 29, 32, 181
pine and, 108
reforestation efforts, 81, 128, 129, 131
reorganization of forestry bureaucracy, 62, 86–87, 124, 180
Seub Nakhasathien and, 33, 34, 35
teak and, 103, 105
territory controlled by, 61
Watershed Conservation Division, 121
See also corruption
Ruhle, George, 157–63, 180
career in US National Parks Service, 169
on conflicts with Indians, 180
in Thailand, 168–71

Sa-ard Boonkird, 96, 115
Salween river basin, 54
Karen farming in, 49
logging operations in, 55, 59, 60, 65, 75
timber output in, 77
Salween National Park, 175
Sanan Kachornprasart, 35
Sanga Sabhasri, 125
Sanitsuda Ekachai, 31, 116, 186
Sarit Thanarat, Field Marshal, 164, 167, 169
Schlich, Wilhelm, 44, 45, 46, 83
Schlich's Manual of Forestry, 45, 48
scientific forestry, 8, 9, 18, 20, 47–48, 183
Brandis and, 43–44, 46, 48
critics of, 48, 49
dual nature of, 108
export of, 147
Germany's influence, 41–43, 144
in India and Burma, 39, 70
industrial plantations and, 99, 142
in Java, 41, 70

Khor Chor Kor and, 32
parallels with wilderness thinking, 180
rules of, 53
Slade and, 49
Seidenfaden, Erik, 92–93, 95, 96, 97
Seidenfaden, Gunnar, 95
Seub Nakhasathien, 12, 32–36
Shell Thailand, 116, 132–35, 136, 137
Siam Rath, 20, 33
Sitthichai Ungphakorn, 185
Slade, Herbert, 8, 38–39, 51, 95, 106
appointment as Siam's first conservator of forests, 56
Bombay Burmah Trading Corporation and, 59–60, 64–65
chao and, 38, 52, 57, 58, 60
at École Forestières, Nancy, 41
European logging companies and, 53, 65
H. A. Slade Building, 78, 94
Indian and Burmese forestry as prototypes used by, 82
on over-cutting, 62, 75–76
Prince Damrong and, 51, 57, 59, 60, 62
questioning use of fire, 49–51, 85
regulation of forest use and, 52–53, 56–59, 60, 65, 86
slash-and-burn agriculture. *See* swidden
Somsak Sukwong, 1, 105, 183
Soon Hua Seng Group, 140
Stebbing, E. B., 41, 46
Stewart Cox, Belinda, 34
Suan Kitti Reforestation Company, 25–28, 29, 116, 121, 132, 136
"clone bank," 28
plantation workers' arrests, 118, 135, 137
See also Kitti Damnerncharnvanich
Surachet Chettamart, 150, 151
swidden, 2, 49

biodiversity and, 74
first pine plantation and, 106
kaingin farmers, 171
teak growth and, 49, 69, 85, 100
Syrach-Larsen, Carl
  author of *Genetics in Silviculture*, 96
  green revolution forestry and, 99–105
  involvement with teak breeding in Thailand, 96–97

taungya, 46, 49, 69, 70, 83–84, 106, 111, 135
teak
  characteristics of, 40, 55, 89, 100, 102, 107
  control over forests of, 52–54
  corruption and, 78–81
  diversity and, 44
  extent of, 56
  extraction of, 71, 75–77, 105
  destruction of, 60
  FIO and, 61
  as foreign exchange earner, 5, 52, 57, 59, 61
  as key species in Thai forestry, 9, 74, 90
  as laboratory for green revolution forestry, 9, 91, 98, 99–105
  laws related to, 67, 123, 124
  logging practices, 68–69, 71, 111, 165
  Nam Chon and, 11, 14, 15
  plantations of, 81–84
  shifting cultivation and, 48, 49, 66, 69, 85
  Slade on fire and, 50
  strategic colonial importance of, 37, 38, 39–41, 52, 63–65, 67
  See also Danish influence on Thai Forestry; Syrach-Larsen, Carl
Teak Improvement Center, 65, 96, 97, 98, 103, 104
Teak Preservation Act, 67

criminalizing common people's use of forests, 67
Team Consulting Co. Ltd., 13
Tem Smitinand, 96
Thai-Japan Reforestation Co., 132, 135–37
Thai Plywood Company, 18, 19
Thap Lan National Park, 178
Thiwa Saphakit, 20
Thung Yai Naresuan Wildlife Sanctuary, 12–16, 18, 34, 177
timber concessions, 9, 60, 61
  cancellation of, 11, 17, 20, 22, 62, 72, 88
  held by European companies, 52, 53, 165
  Karen men working in, 68
  rules for, 19, 67
Tottenham, W. F. L., 59, 67, 75

Udall, Stewart, 169
UNESCO World Heritage Site, 11
  Huai Kha Khaeng, 13, 34
  Dong Phayayen-Khao Yai Forest Complex, 178
U-nite village, 15, 16
Uthit Kud-In, 21

Veerawat Dheeraprasart, 11, 89, 184
von Langen, J. G., 97

Wat Hua Nam Phut, 30
Watershed Conservation Division. *See under* Royal Forestry Department (RFD)
Wiboon Khemchalerm, 140
wilderness ideal, 10, 147–50, 183
  Glacier and, 159, 162
  Muir in Yosemite, 152
  as model for Thai national parks, 180
  US wilderness as *emptied* land, 157
  wilderness advocates, 168

    Yellowstone and, 154
wildlife
    Blackfeet Indians and, 159
    Boonsong Lekagul on, 164–67
    conservation, 18, 166, 188
    impact of road inside Khao Yai forest complex and, 178
    Seub Nakhasathien on, 32–36
    Shell's claims about eucalyptus, 133
    US parks authorities' approach to, 160–62
Wildlife Conservation Division, 16, 35, 147, 171
Wildlife Conservation Law, 19
wildlife sanctuaries, 3, 7, 19
    based on core belief of no human disturbance, 147, 171
    community forests and, 25
    expansion of, 174, 176, 180
    inhabitants of, 151
    Nam Chon and, 12, 13
    relics of Cold War, 163
Winai Subrungruang, 73, 84
World Wildlife Fund, 168

Yam Malla, 186
Yellowstone, 10, 150, 154–57, 169, 181
Yosemite, 10, 150, 151, 152–54, 169, 181

Zackrisson, Olle, 142, 145, 146, 187